ENVIRONMENT INSTITUTE

MODELLING LAKE POLLUTION

Giovanni Rossi

JOINT RESEARCH CENTRE

EUR 13998 EN

COMMISSION OF THE EUROPEAN COMMUNITIES

1991

Published by the
COMMISSION OF THE EUROPEAN COMMUNITIES
Directorate-General
Telecommunications, Information Industries and Innovation
Batiment Jean Monnet
L-2985 LUXEMBOURG

LEGAL NOTICE

Cataloguing data can be found at the end of this publication.

Luxembourg: Office for Official Publications of the European Communities, 1991
ISBN 92-826-3267-9 Catalogue number: CD-NA-13998-EN-C

Printed in Italy

Contents

Preface

Interdisciplinary studies on freshwater ecosystems, which were carried out at the Ispra Establishment of the Joint Research Centre of the Commission of the European Communities as activities of the project Water Quality (Programme Protection of the Environment), resulted in a series of scientific publications concerning various lakes of the subalpine region and various types of pollution.Most of these studies followed a modelling approach, in the sense that field and laboratory experimental research was planned and results were interpreted with a mathematical model as a reference point.

This monograph is the result of a long interaction between modeller and experimentalist in tackling freshwater ecology problems. Its purpose is twofold: to present the concepts underlying lake modelling in a simple and concise way and to discuss a selection of lake pollution cases, as a synthesis of a pluriannual activity in which both interpretative and predictive modelling were developed.

Part I is a revised and extended version of the author's article "Modeling Lake Eutrophication" in the Encyclopedia of Fluid Mechanics, vol.10, Surface and Groundwater Flow Phenomena, N.P. Cherimisinoff ed., Gulf Publishing Comp., Houston.

Part III is based on papers published by the author in international journals, whose reference is given at the end of the respective chapter.

At the end of each chapter a list of titles is given, which are suggested to the reader for further development of the subject concerned.

Furthermore, the following titles are suggested as comprehensive reviews:

A. Lerman (ed.), Lakes: Chemistry, Geology, Physics. Springer, New York, 1978 (new edition in press)

G.T. Orlob (ed.), Mathematical Modeling of Water Quality: Streams, Lakes and Reservoirs, J. Wiley & Sons, New York, 1983.

W. Stumm (ed.), Chemical Processes in Lakes, ibidem, 1984.

Acknowledgments

The author is indebted to the experimental staff of the Aquatic Ecology and Heavy Metal Chemistry Laboratories of J.R.C. Ispra, and in particular to the respective heads, Guido Premazzi and Herbert Muntau, for the valuable collaboration during the project work, as well as to Johanna Sangregorio Hartman, who has carried out computer programming for the various models.

Support and encouragement by the managers of the Programme Protection of the Environment Friedrich Geiss and Bruno Versino to the Water Quality project activities and to realize this publication were greatly appreciated.

Introduction

The mathematical models discussed in the following chapters are devoted to the description of a lake as a whole. Pollution episodes, affecting a small portion of the lake surface (e.g. a small coastal zone) during a limited time interval, are only occasionally considered.

As a matter of fact, mixing (e.g. of the *coastal* water with the *pelagic* one) and chemical reactions tend to reduce the pollutant concentration in the region affected within a time which can be considered as short with respect to the characteristic evolution time of the whole lake.

As a consequence such events, although relevant from the health point of view sometimes, modify the whole lake state to a negligible extent.

The **state of the lake** is determined by the distribution of various abiotic chemical substances, which are dissolved in water or aggregated to suspended matter, as well as by the evolution of biologic populations living in the water body.

The most important of the latter ones is the algal population (phytoplankton, microphytes). Among abiotic substances a particular role is played by algal nutrients (e.g. phosphorus and nitrogen compounds) and by oxygen, which is necessary to the life of aquatic animals.

As for each chemical substance, which is present in the water body, a lake is characterized by a "natural" concentration, as determined by its natural environment and by populations living inside, a concentration which corresponds to a equilibrium condition unperturbed by human activity.

Lake pollution is caused by the presence of a chemical substance in the lake at a concentration greater than the natural one. When this substance is an algal nutrient lake pollution is termed **eutrophication**, while the general term **chemical pollution** is adopted for substances which don't enter or enter in very small quantities the metabolism of aquatic plants (e.g. heavy metals).

Effects of eutrophication are high concentration of algal biomass (high turbidity, low transparence) in the surface water, high sedimentation fluxes of organic detritus, and oxygen depletion in deep water because of consumption for organic detritus mineralization by bacterial activity.

Chemical pollution affects biologic populations to a lesser or greater extent, when pollutant concentration exceeds a toxicity threshold, as determined by ecotoxicologic investigation.

The two forms of lake pollution are concerned by mathematical models which have many common features, beginning with the description of the physical support, i.e. with lake thermo-hydrodynamics concepts.

As a matter of fact, the fundamental equation of the model is a classical mass conservation equation, whose general form is

$$\partial C/\partial t = -\operatorname{div}(\vec{v}\,C) + S\,,$$

where C is the concentration of the substance considered, S includes source and interaction terms, and $\vec{v}$ is the displacement velocity of the "particles" of this substance.

The velocity $\vec{v}$ is obviously connected with water velocity, as given by a separate thermo-hydrodynamic model.

A lake pollution (ecological) model is mainly concerned with biologic and chemical processes, and has wide recourse to averaging procedures as for space (especially horizontal x,y coordinates) dependence. For this reason the recourse to a separate hydrodynamic model is generally limited to the strict necessary.

The **fundamental distinction** between a lake eutrophication model and the distribution model of a chemical pollutant in a lake is the different connection between the concentration of abiotic substances and the evolution of biotic variables.

Algal biomass and nutrient concentration are tightly connected, in the sense that their conservation equations have to be simultaneously solved, while in a chemical pollutant distribution model, which neglects toxic effects on the biota, algal cells are only one of the particle species to which the pollutant can be aggregated.

In other words, a separate (trophic state) model supplies the biomass concentration as a input datum to the pollutant distribution model. In the trophic state model the toxic effects of the chemical pollutant are evaluated by introducing into the algal biomass evolution equation a dependence on the pollutant concentration, which was separately calculated.

Also the time scales for the two types of model are generally different. Algal dynamics has characteristic times of the order of days, while the concentration of chemical pollutants, if they are not particularly reactive, varies much more slowly, especially when average values in a large lake region are envisaged. Only a long term predictive trophic state model, which is mainly concerned with total phosphorus concentration, is allowed to use the same mathematical formalism as a chemical pollution model.

From a scientific point of view the specific interest of ecosystem modelling resides in the attempt at a quantitative description of the interaction between biologic populations and the natural environment (here "natural environment" is opposed to "laboratory conditions"). In the case of lacustrine ecosystems, even neglecting trophic levels higher than phytoplankton, this is a very hard job, and practical purposes, which impose drastic simplifications, are generally prevailing.

Then the modeller's skill is to identify the essential features of the experimental framework and to draw correct consequences from the comparison between experimental and model results.

Part I

MODELLING LAKE EUTROPHICATION

1. Phenomenology of Lake Eutrophication

The term "eutrophication" indicates deterioration of the aquatic environment due to excessive fertilization of natural waters (lakes, reservoirs, rivers and marine coastal zones), i.e. excessive increase of the algal biomass and its consequences, such as reduced transparency of surface waters and oxygen depletion in deep waters.

These phenomena interfere with use of water by man, in particular with drinking water quality, and for this reason eutrophication has emerged as one of the most significant environmental problems in the modern age.

There is a natural and a cultural eutrophication process. Natural eutrophication is a normal ageing process of lakes, caused by transportation of material from the catchment area into the water body.

Cultural eutrophication is caused by an excessive input to natural waters of aquatic plant nutrients as a consequence of man's activities. It is linked to demographic, agricultural and industrial development, not assisted by a proper territorial planning of drainage basins.

The aquatic environment, which is most directly involved in cultural eutrophication processes, is that characterized by the slowest water renewal, i.e. the lacustrine one. As a matter of fact, almost all lakes in the densely populated and industrial countries are affected by eutrophication to a greater or lesser extent.

Although various chemical elements (C, O, H, N, P, Si, S, Mg, Ca + oligo elements) are utilized by aquatic plants in the photosynthetic process, phosphorus is considered as the most important ecologically, because the ratio of phosphorus to other elements in aquatic as in other living organisms is greater than in their environment, and a deficiency of phosphorus tends to limit their productivity. Moreover it is both technically and economically the most feasible to control.

For this reason, attempts at a quantitative description of the lake eutrophication process, as well as at a prediction of its evolution, by means of mathematical models, are primarily concerned with the phosphorus cycle and mass balance in the water body.

The basis of all macroscopic phenomena which take place in a lake and are relevant to the eutrophication process, is, at the level of microscopic events, *the photosynthetic reaction*, which transforms inorganic phosphorus, dissolved in water as the orthophosphate ion $(PO_4)^{3-}$, into a component of the algal cell protoplasm.

This reaction, which utilizes the energy of solar radiation by means of algal pigments (e.g. chlorophyll-a), is described by the following stoichiometric equation:

$$106CO_2+16NO_3^-+HPO_4^{--}+18H^++122H_2O \rightleftharpoons C_{106}H_{263}O_{110}N_{16}P+138O_2.$$

The inspection of this equation allows us to determine a quantity, which is relevant to the assessment of the trophic stage of a water body, i.e. the ratio between oxygen produced (consumed) and phosphorus consumed (produced) (~ 140 mg O_2/mg P).

Moreover the formula adopted for the algal protoplasm exhibits the values of the mass ratios between its essential components (C, N, P).

Observable **macroscopic phenomena** relevant to lake eutrophication are related both to external actions on the lake (solar radiation, precipitation, wind action, water inflow-outflow associated with transport of matter into and away from the water body, release of substances from the lake walls) and to physical, chemical and biological processes, which take place within the lake water. Internal processes include:

- horizontal and vertical advective and diffusive transport of dissolved substances;
- sedimentation of particulate matter;
- uptake of dissolved nutrients (in particular orthophosphate phosphorus P–PO_4) by algal cells;
- increase of algal biomass and oxygen production as consequences of the photosynthetic reaction;
- transformation of living biomass into organic detritus as a consequence of the ageing and death of algal cells;
- regeneration of nutrients (in particular P-PO_4) due to bacterial activity on organic detritus and associated oxygen consumption.

Solar radiation in its daily and annual evolution and wind action in its periodic or episodic occurrences are at the origin of thermal stratification and destratification of lake water bodies. Wind action on warmer surface water and contrasting buoyancy forces create a layer of maximum water temperature and density gradients (thermocline), which separates a well mixed surface layer (epilimnion) from the deeper ones (hypolimnion). After thermal stratification disappears during winter, wind action causes the lake water to overturn. This phenomenon can affect the deepest layers (holomictic lakes), or a certain depth varying from one year to the next, and only occasionally the lake bottom (meromictic lakes).

Experimental investigation on deep thermally stratified lakes has revealed two types of *phosphorus cycles.* The cycle which takes place in the epilimnion (trophogenic zone) consists of the uptake of inorganic phosphorus (P-PO_4) by algae, its assimilation as algal phosphorus, transformation of algal into organic particulate phosphorus associated with organic detritus (dead algae), and finally regeneration of orthophosphate phosphorus by bacterial activity.

A second type of phosphorus cycle concerns the whole water column and involves sedimenting particulate phosphorus and orthophosphate which is regenerated at the water-sediment interface or even in the highest sediment layers, from which it is released in particular physical and chemical (e.g. anoxic) conditions, and eventually reaches the photic zone again, where it is again used for algal growth (fig.1). In shallow lakes the two cycles take place in the same lake waters and released orthophosphate is promptly available to algae. Lakes which are becoming eutrophic are characterized by the increasing weight in the phosphorus mass balance of the phosphorus cycle going through the sediments, as a consequence of intense organic detritus production and sedimentation and inherent oxygen consumption.

All this phenomenology is experimentally investigated on the basis of lake water sampling at different points of the lake surface, at various depths and at various times during the annual evolution.

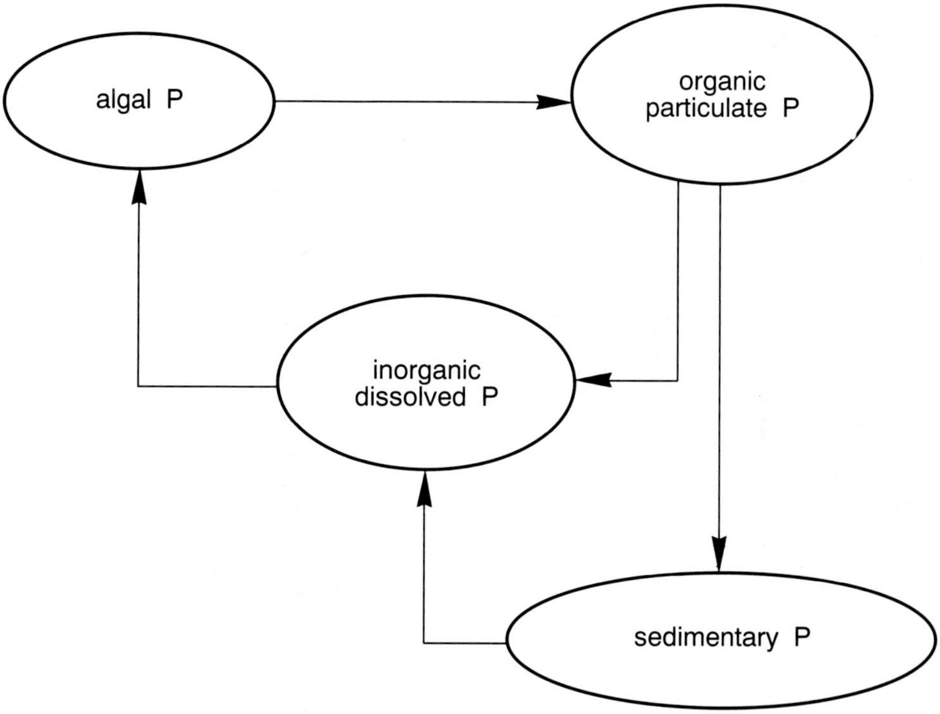

Fig. 1 Outline of the phosphorus cycles in a lake: the "direct" cycle through mineralization of the particulate organic phosphorus distributed over the water column, and the cycle involving sedimented particulate phosphorus which is released as dissolved inorganic after mineralization of the organic fraction in the sediment.

Table I presents a list of physical, chemical and biological variables, which must be systematically measured to control the lake trophic evolution. In the following chapters will appear that not all these variables are explicitly considered by lake eutrophication models.

Annual evolution of chlorophyll-a, orthophosphate phosphorus and dissolved oxygen concentration (fig. 2) reveal the anomalies of an eutrophic lake.

The experimental framework presented above does not include the *zoo-planktonic component* of lake biomass. Its effect on the evolution of algal biomass will be incorporated into the daily "loss" coefficient of living algal biomass (losses due to

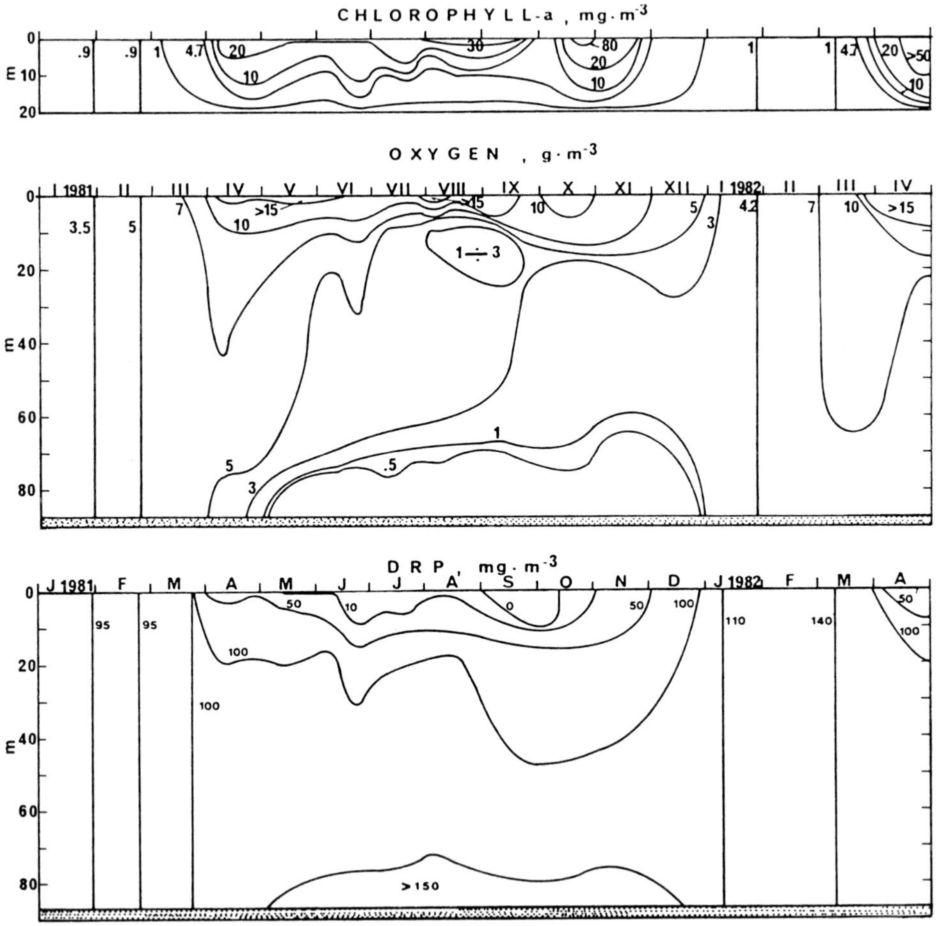

Fig. 2 Depth-time diagrams of chlorophyll-a, dissolved oxygen and dissolved inorganic (reactive) phosphorus concentrations in a eutrophic lake (Lake Lugano).

grazing of algae by zoo-plankton). This simplification is often adopted in global eutrophication models.

Finally, external *nutrient (phosphorus) loading* to the lake, whose correct assessment is of primary importance for all management purposes, is usually associated with the measured water inflow rate through an "equivalent" external nutrient concentration in the inflowing water, which must include the contributions from diffuse sources (runoff, etc.).

Table I List of fundamental limnological variables considered in O.C.D.E. Projects for the Surveillance of Inland Water Eutrophication

1.	**Physical parameters** Temperature, electrical conductivity, transparency, solar radiation
2.	**Chemical parameters** pH, dissolved O_2; P and N compounds; SiO_2; alcalinity and acidity; Ca^{++} and Mg^{++}; Na^{+} and K^{+}; SO_4^{--} and Cl^{-}; total Fe.
3.	**Biological parameters** Phytoplankton (chlorophyll - a); primary production; organic C.

Sources and suggestions for further reading

Eutrophication of Waters: Monitoring, Assessment and Control.
O.C.D.E., Paris, 1982, pp.154.

2. The Lake as a Mixed Reactor

All mathematical models describe some idealization of the reality by means of differential equations. Although physical lake modelling must be as old as hydrodynamics, or at least as physical oceanograpy, ecological (global) lake modelling is rather recent. Because of complexity of the phenomenology outlined in the previous chapter, ecological lake modelling has historically started from a drastic idealization: the lake as a mixed reactor, where various chemical forms of the limiting nutrient (phosphorus) are transformed one into the other, the total concentration being determined by exchanges across the boundaries (inflow-outflow, sedimentation-release).

Such an idealization disregards the details of internal processes, in particular algal growth and time evolution of algal biomass, nevertheless it has always been considered a useful tool for discussions on the long term evolution of eutrophicated lakes. The model consists of a single differential equation for the concentration of total phosphorus, transported into the lake by inflowing water, homogeneously distributed throughout the lake water body, and lost from this through outflow and sedimentation:

$$V\, dC/dt = q_{in}C_{ext} - (q_{out}/V + \sigma)VC \qquad C(0) = C_o \tag{1}$$

where:

V	≡	lake volume;
C, C_{ext}	≡	concentration of total P in lake and in the inflowing water, respectively;
q_{in}, q_{out}	≡	inflow and outflow rates;
σ	≡	"net" sedimentation rate (i.e. sedimentation minus release).

If q_{in}, q_{out}, C_{ext}, σ, V are assumed as constant (average values in the time interval considered), the solution of (1) is:

$$C = C_0 \exp\left[-\left(\frac{q_{out}}{V} + \sigma\right)t\right] + \left[1 - \exp\left(-\left(\frac{q_{out}}{V} + \sigma\right)t\right)\right] \frac{\frac{q_{in}}{V} C_{ext}}{\frac{q_{out}}{V} + \sigma}. \tag{2}$$

The inverse of the characteristic time of the transient term in Eq.(2):

$$\frac{1}{\tau} = \frac{q_{out}}{V} + \sigma \tag{3}$$

is the sum of the inverse of the water mean residenc time $\tau_w = V/q_{out}$ and of the inverse of the characteristic time of the "net" or "apparent" sedimentation process $\tau_{sed} = \sigma^{-1}$.

The steady state concentration of total phosphorus in lake:

$$C_\infty = \frac{q_{in}}{V} C_{ext} \Big/ \left(\frac{q_{out}}{V} + \sigma \right), \tag{4}$$

gives a relationship between phosphorus loading and in-lake concentration, which has been widely used for eutrophic lake classification. Since V = Sh, where S is the surface area and h the mean depth of the lake, and introducing the *specific phosphorus loading* $L = q_{in}C_{ext}/S$, Eq.(4) becomes:

$$C_\infty = L \Big/ \left(\frac{h}{\tau_w} + \frac{h}{\tau_{sed}} \right). \tag{5}$$

On the basis of experimental observations on a large number of lakes, the ratio h/τ_{sed} has been considered as approximately independent of lake morphology and taken as 10 m yr–1. Hence the relationship between specific phosphorus loading and in-lake steady state phosphorus concentration only involves the ratio h/τ_w, which appears as the so-called "*hydraulic loading to the lake*" in the original model equation (1) ($h/\tau_w = q_{in}/S = q_s$). In a plane ($h/\tau_w$,L) the iso-concentration lines are straight lines of equation

$$L = \left(\frac{h}{\tau_w} + 10 \right) C_\infty . \tag{6}$$

If a critical steady state phosphorus concentration is defined, above which the lake must be considered as eutrophic, the corresponding straight line divides the plane (h/τ_w,L) in two regions, that of eutrophic lakes above the critical straight line, and that of meso-oligotrophic lakes below it.

Further modifications to Eq.(6) have been introduced to show the interdependence between L and h/τ_w and τ_w separately, and to agree with the non linearity in the dependence of L on τ_w exhibited by experimental observations. These modifications are empirical in nature and dimensionally not consistent. On the basis of a statistical correlation between τ_{sed} and τ_w, the following version of eq. (6) was finally proposed:

$$L/q_s = C'_\infty (1 + \sqrt{\tau_w}), \tag{7}$$

where L/q_s is the average phosphorus concentration in inflowing waters.

The ratio between the average phosphorus concentration in inflowing waters and in lake is presented in fig.1 as a function of τ_W, following Eq.(6) and Eq.(7). From the determination of this ratio for a lake of known morpho-hydrologic characteristics it is easy to deduce the external phosphorus loading compatible with a desired in-lake concentration.

The main limitation to the use of Eq.(1), and consequently of Eq.(6), as a tool for the prediction of eutrophic lake evolution is connected with the *sedimentation term.*

Gross sedimentation concerns primarily the particulate fraction of phosphorus; furthermore it is experimentally known that, after a certain time in a progressive eutrophication process, the summer level of algal biomass, which gives the main contribution to the sedimentation flux, as well as the phosphorus release flux from sediments, hence net sedimentation, are approximately independent of in-lake (dissolved) phosphorus concentration.

As a consequence, the sedimentation term in Eq.(1) must be considered as constant, and the characteristic time increased to τ_W. Predictions on lake evolution after reduction of external phosphorus loading require a proper evaluation of the sediment dynamics and the deriving phosphorus release (see chs. 4 and 7).

Finally, a mixed reactor model does not consider the fact that, during a considerable period in the year, the deepest layers of the water body (hypolimnion) are excluded

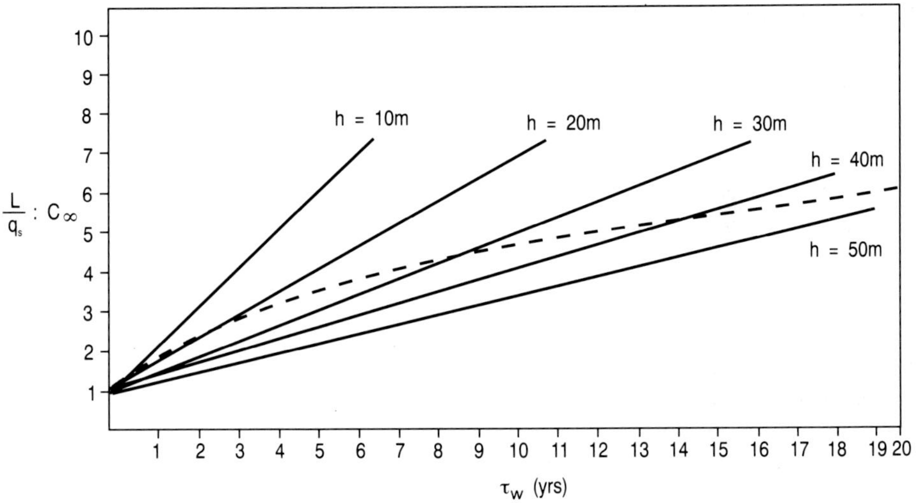

Fig. 1 The ratio between total phosphorus concentrations in inflowing water and in lake as a function of water mean residence time for various lake mean depths (straight lines, Eq.(6)). The dashed curve represents the function $y = 1 + \sqrt{x}$ (Eq.(7)) which is valid for all mean depths.

from the lake water renewal process. Replacing τ_W with the effective water mean residence time $\tau_W' > \tau_W$ in Eq.(7) written in the form

$$L = C_\infty \frac{h}{\tau_W} (1 + \sqrt{\tau_w}) \qquad (8)$$

has the consequence of lowering the critical loading.

The connection between phosphorus in-lake concentration and phytoplankton biomass concentration is extraneous to the model summarized by Eq.(1). On purely empirical (statistical) grounds, OCDE suggested the following relationship:

$$\overline{CHL} = a\bar{P}^b, \qquad (9)$$

where $\overline{CHL}$, $\bar{P}$ are the average annual concentrations of chlorophyll-a and total phosphorus, respectively, and the values of a and b, obtained from statistical treatment of various sets of data, lie in the intervals: $0.08 \leq a \leq 0.52$; $0.81 \leq b \leq 1.27$.

Sources and suggestions for further reading

1. R.A. Vollenweider, Advances in defining critical loading levels for phosphorus in lake eutrophication.
   Mem. Ist. Ital. Idrobiol., Vol. 33, 53-83 (1976)
2. R. Gächter and D.M. Imboden, Lake restoration.
   In “Chemical Processes in Lakes” (W. Stumm, ed.) J. Wiley & Sons, New York, pp. 365-388 (1984)

3. The Evaluation of Water Mean Residence Time in a Lake

As already appears in the framework of the simplest conceivable eutrophication model as outlined in ch.2, the mean residence time of water in a lake is a quantity of fundamental importance in predicting the evolution of lake water quality.

When this concept is applied to an arbitrary water volume, this parameter summarizes the water exchange properties of this region with respect to the surrounding one. In fact, a rigorous definition of water residence time in a volume V, implies the description of the water velocity field within the volume V and along the surface S enclosing V.

A practical calculation of water mean residence time was performed by Piontelli and Tonolli for the meromictic Lake Maggiore in Northern Italy, utilizing the statistical information on lake water behaviour such as thermal stratification and destratification, and water overturn historical data. Their procedure was based on the following assumptions: (a) the inflowing waters are homogeneously mixed with the lake water above the thermocline (epilimnion); (b) the outflowing water comes from the epilimnion only. Starting from these assumptions a formal mathematical procedure can be established, which is easily adapted to computer calculations.

In meromictic lakes the winter water overturn concerns only a part of the lake volume (the *mixolimnion*), while it concerns the total lake volume only in particular meteorological conditions (very cold and windy winters). On the basis of historical information it is possible to identify the average thickness of mixolimnion, as well as the average length of the time interval between two consecutive total water overturns.

At time t the lake water volume above the thermocline (epilimnion volume V_{ep}) is partially occupied by water already present in the lake at time t_o when the description of the water renewal process began. If $\Omega(t)$ is the volume occupied by original water in the epilimnion at time t, then clearly $\Omega(t_o) = V_{ep}$, and the dilution coefficient of original water at time t is $\Omega(t)/V_{ep}$. If precipitation and evaporation are neglected, then $\Omega(t)$ varies only as a consequence of inflow and outflow, and the epilimnion volume $V_{ep}(t)$ varies as a consequence of the variation in thermocline depth. Thermocline depth and volume V_{ep} can be taken as stepwise functions of t. If it is assumed that during the time interval $(t,t+\delta t)$ the inflowing water and original water become homogeneous, the variation of $\Omega(t)$ during the interval is given by:

$$\delta\Omega = -q\,(t)\,\frac{\Omega(t)}{V_{ep}(t)}\,\delta\,t \tag{1}$$

where $q(t)$ is the outflow rate (equal to the inflow rate) at time t. Then, during the time interval (t_o,t_1) in which the epilimnion volume is constant, the volume of original water $\Omega(t)$ is obtained from the solution of the following differential equation:

$$d/dt\,\Omega = -\frac{q\,(t)}{V_{ep}}\,\Omega\,(t) \qquad \Omega\,(t_o) = \Omega_o = V_{ep} \tag{2}$$

If the annual cycle is divided into n intervals of constant epilimnion volume $V_{ep,i}$ ($i = 1,...,n$), the solution of Eq.(2) in the interval i is given by:

$$\Omega_i\,(t) = \Omega_{io}\,\exp\left[-\frac{1}{V_{ep,i}}\int_{t_i}^{t} q_i\,(t')\,dt'\right] \qquad \Omega_{io} = \Omega_i\,(t_i)\;(i = 1,\ldots,n) \tag{3}$$

The assumption of instantaneous homogenization of the inflowing water with the "old" one leads to underevaluation of the original water volume flowing out from the lake, and consequently to overestimation of the residence time. As a matter of fact, the water renewal process depends on the location of inlets and outlets, in such a way that the instantaneous homogenization assumption can be particularly crude in certain cases.

Because of large fluctuation in the inflow rate, the time interval of constant epilimnion volume $V_{ep,i}$ must be subdivided in subintervals $\Delta t_{i,k}$ of constant inflow rate $q_{i,k}$; then Eq.(3) gives:

$$\Omega_i\,(t_i + \Delta t_i) = \Omega_{io}\exp\left(-\frac{1}{V_{ep,i}}\,\Sigma_k\,q_{i,k}\Delta t_{i,k}\right). \tag{4}$$

The numerical procedure for meromictic lakes starts from the definition of the "normal" annual cycle (i.e. a cycle with partial water overturn at the end of winter). Proper utilization of Eq.(4), taking account of the particular situation which occurs when the epilimnion volume $V_{ep,i}$ is smaller (stratification) or greater (destratification) than $V_{ep,i-1}$, allows us to evaluate the dilution coefficient of original water in the mixolimnion at the end of a normal cycle:

$$\alpha_{mix} = \alpha_{ep,n} = \frac{\Omega_n (t_n)}{V_{ep,n}} \quad (5)$$

where $V_{ep,n}$ is the volume of the mixolimnion.

After m identical normal annual cycles the dilution coefficient of the original water in the mixolimnion is $(\alpha_{mix})^m$. The $(m+1)^{th}$ annual cycle, characterized by total water overturn, is treated in an anologous way, starting with the initial condition $\Omega_o = (\alpha_{mix})^m.V_{ep,1}$.

The value of the dilution coefficient of the original water at the end of the pluriannual cycle of T = m+1 years refers to the whole lake, homogeneously mixed. The mean residence time of the lake water is obtained as a function of this dilution coefficient (α) in the following way.

An averaging process must be performed, weighting the mean residence time of water outflowing during the nth cycle of T years:

$$\tau_n = (nT + (n-1)T)/2 = (2n-1)T/2 \quad (6)$$

with the fraction of the total lake volume flowing out during the same cycle $p_n = \alpha^{n-1} - \alpha^n = \alpha^{n-1} (1-\alpha)$:

$$\tau = \sum_{1}^{\infty}{}_n\, p_n \tau_n = (1-\alpha) \sum_{1}^{\infty}{}_n (2n-1)\, \alpha^{n-1} . T/2 \quad (7)$$

After summation of the series the following expression is obtained:

$$\tau = \frac{T}{2} \cdot \frac{1+\alpha}{1-\alpha} \quad (8)$$

If total water overturn occurs every year (m = 0), the calculation gives α and Eq.(8) $\tau = (1+\alpha) / [2 (1-\alpha)]$ directly.

The method was applied to Lake Maggiore ($V = 37500x10^6\ m^3$; $q_{out} = 9272x10^6\ m^3yr^{-1}$; T = 6 years) with the following results:

$$\alpha_{mix} = 0.808 \qquad \alpha = 0.665 \qquad \tau = 14.9\ yr,$$

to be compared with the "theoretical" water mean residence time $V/q_{out} = 4.04$ yr. The water mean residence time in the mixolimnion of Lake Maggiore (average depth = 192 m in the period 1963-70) is given by $\tau_{mix} = (1+\alpha_{mix}) / [2 (1-\alpha_{mix})] = 4.71$ yr.

Morphologically complex lakes (as for instance Lake Lugano and Lake Como in Northern Italy) present sharp heterogeneities in the mean residence time of water in different basins of the lake, which are due to the distribution of inlets and outlets, and to the holomictic or meromictic behaviour of each separate basin. In these cases the evaluation of the mean residence time of water has an ecological significance only for single basins, and should be supported by hydrodynamic calculations.

Experimental information on water residence time can be obtained by the use of natural or artificial tracers, i.e. substances whose concentration can be measured at the inlets at time t_0, and whose detectable presence at the point $\vec{r}$ in the lake and in particular at the outlet at time $t > t_0$, is due only to water transport processes. As a matter of fact, the definition (7) of water mean residence is equivalent to the following one:

$$\tau = \int_{t'}^{\infty} [1 - A\,(t-t')]\,dt \tag{9}$$

where A (t-t') is the probability that a water parcel inflowing into the lake at time t' has already left the lake at time t. Complementary experimental information on water residence time concerns the time which elapsed from when a water parcel was in the surface mixed layer, i.e. when the ^{3}He isotope contained in lake water was in equilibrium with atmospheric ^{3}He. Mass spectrometric determination of ^{3}He excess, due to tritium decay, in lake water at depth z, allows us to evaluate the "age" of that water parcel (age = 0 characterizing water in the mixed layer).

This technique was made possible by the high tritium concentrations existing in the atmospheric precipitations after the atomic bomb explosions.

Sources and suggestions for further reading

1. R. Piontelli e V. Tonolli, Il tempo medio di resistenza delle acque lacustri. Mem. Ist. Ital. Idrobiol., Vol. 17, 247-266 (1964)
2. G. Rossi, et al., On the calculation of the mean residence time in monomictic lakes, Hydrol. Sciences Bull., XX, 4-12 (1975)
3. T. Torgersen, et al., A new method for physical limonology: tritium-helium 3 ages results for Lakes Erie, Huron and Ontario. Limnol. Oceanogr., 22, 181-193 (1977)

4. Modelling Lake Sediment Interaction

The considerations presented in ch.2 allow us to conclude that even the simplest lake eutrophication model should include the evaluation of the effect of sediment interaction on the lake phosphorus budget.

In reality this interaction results from various complex chemical processes involving phosphorus compounds of different solubility, after mineralization of organic phosphorus associated with sedimented matter. Bacterial activity and chemical reaction rates are dependent on physico-chemical parameters such as water temperature, pH and redox potential. Moreover, remobilization of sedimented matter is caused by physical and biotic agents (water turbulence, bentonic plankton, ...), which are even less easy to model.

At the same level of semi-empirical approximation as the mixed reactor model discussed in ch.2, a two-box lake-sediment model can be envisaged to evaluate the effect of sediment activity on the evolution of an eutrophic lake: it represents the state of lake and sediment, each by means of a single variable.

Having recourse to suitable assumptions as for sediment processes (fig. 1) and adopting time-independent (average) coefficients, the resulting system of differential equations can be solved analytically.

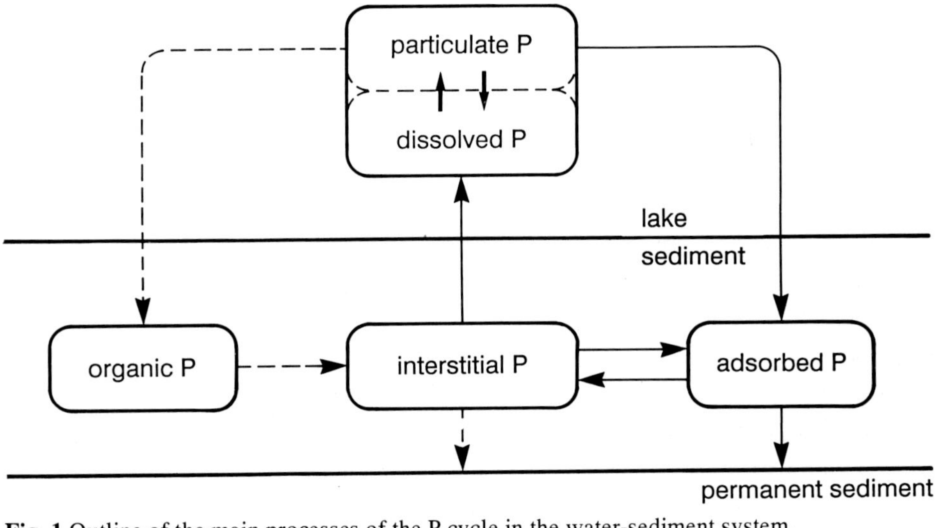

Fig. 1 Outline of the main processes of the P cycle in the water-sediment system.

In order to characterize the lake state only by means of the average total P concentration P_L (t) (mg m^{-3}), it is necessary to represent outflow and sedimentation flux in terms of P_L, i.e. to insert suitable factors into the related equation coefficients. These factors are the ratio β between average P concentrations in the epilimnion and in the whole lake, and the ratio π defined as the particulate fraction of the average total phosphorus in-lake concentration. This is equivalent to the assumption that the P concentration in suspended matter is proportional to the dissolved phosphorus concentration, i.e. that the sedimentation flux at the lake-sediment interface is mainly controlled by abiotic processes.

Then the balance equation for total phosphorus in the lake is as follows:

$$V_L \frac{dP_L}{dt} = \underbrace{- q_{out}\, \beta P_L}_{\text{outflow}} \underbrace{- v_s\, A\, \pi P_L}_{\text{sediment.}} + \underbrace{q_{in}\, P_{ext}}_{\text{inflow}} + \underbrace{A_s\, \varphi_r}_{\text{release from sediment}} \tag{1}$$

where:

v_s	=	= is the settling velocity of particulate matter (m yr^{-1});
P_{ext}	=	= the total P concentration in inflowing water (mg m^{-3});
V_L	=	= lake volume (m^3);
A	=	= lake surface area (m^2);
A_S	=	= area of active sediment surface;
φ_r	=	= P release flux from sediments (mg m^{-2} yr^{-1}).

Putting:

$$\frac{q_{out}\beta}{V_L} = k_{out}\ ; \quad \frac{v_s A \pi}{V_L} = k_{sed}\ ; \quad \frac{q_{in}}{V_L} = k_{in}\ ; \quad \frac{V_L}{A} = \bar{z}$$

eq. (1) becomes:

$$\frac{dP_L}{dt} = -(k_{out} + k_{sed})\, P_L + k_{in}\, P_{ext} + \frac{\varphi_r}{\bar{z}}\, .$$

The flux φ_r has to be expressed in terms of the variable which characterizes the state of the sediment.

In general, sediment cores from eutrophic lake exhibit a surface layer, where total P concentration varies irregularly with depth and is, on average, significantly higher than in deeper ones. This layer, whose thickness h_s is if order 0.1 m or even of few cm in deep lakes, is affected by turbulent mixing and bioturbation. In the following, reference is made to it as to the *active sediment layer*, and to the layers below it as to the *permanent sediment*

If r_s is the *sedimentation rate* of the lake (g m^{-2} yr^{-1}), ρ_s the average density of solid sediment (g m^{-3}), and p the average porosity of surface sediment, a characteristic time (renewal time of the active layer):

$$\tau_s = \frac{h_s \ \rho_s \ (1-p)}{r_s}$$

can be defined and evaluated as an intrinsic property of the sediment.

The choice of the state variable for the active sediment layer is based on the following assumptions.

1) The mean residence time of phosphorus in the exchangeable organic phosphorus compartment is negligible with respect to τ_s. Then the whole sedimentation is assumed to feed directly the inorganic phosphorus compartment (dissolved in pore-water + adsorbed on solid matter). Hypolimnetic anaerobic conditions can slow down this feeding process during summer stagnation or even determine the formation of layer of organic detritus at the water-sediment interface. However this layer is rapidly demolished by the bacterial activity as soon as aerobic conditions are re-established.

2) P exchange between pore water and solid sediment is a rapid process: equilibrium is established in a time negligible with respect to τ_s. Then a quantitative relationship (*adsorption isotherm*) is assumed to connect P concentration in pore water P_{int} (mg m^{-3}) and P concentration in solid matter P_{ads} (mg g^{-1}).

This relationship is assumed to be linear for $P_{int} < P_{int,s}$:

$$P_{ads} - P_{ads,o} = a\,P_{int} \qquad (a = m^3\,g^{-1}) \tag{2}$$

where $P_{ads,o}$ has to be interpreted as the P concentration in solid sediment which is not exchangeable.

For $P_{int} \geq P_{int,s}$ solid sediment is assumed to be saturated by phosphorus, then:

$$P_{ads} = \text{const.} = P_{ads,s} = a\,P_{int,s} + P_{ads,o}$$

The values of a, $P_{int,s}$, $P_{ads,o}$ are drawn from experiments on sediment cores (fig. 2).

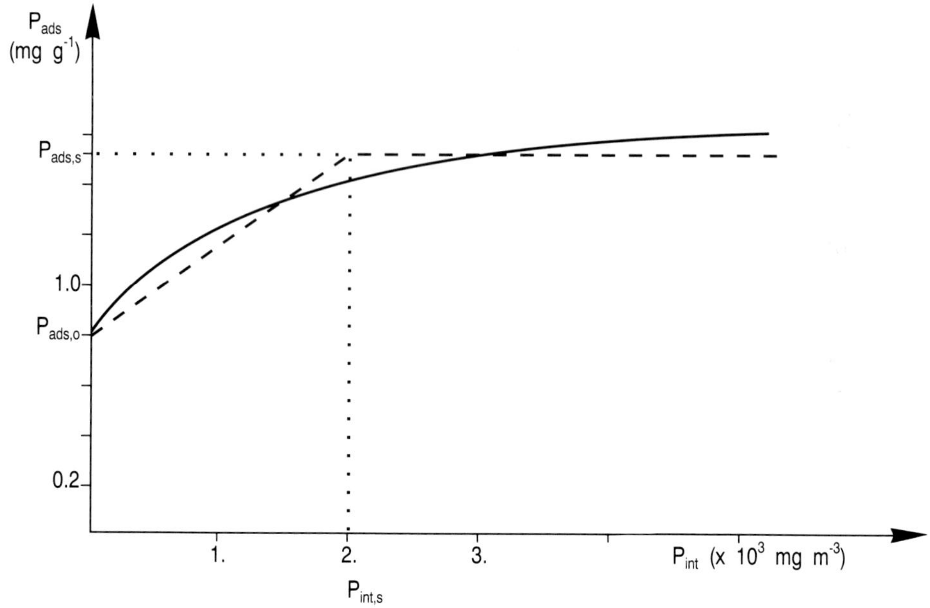

Fig. 2 P adsorption isotherm (—) and its linear approximation as adopted in the proposed model (---).

Referring us to the inorganic compartment of sedimentary P, the mass balance equation is as follows:

$$\frac{d}{dt}\left(p\, V_s\, P_{int} + MP_{ads}\right) = \left(\frac{pV_s}{a} + M\right) d\, \frac{P_{ads}}{dt} =$$

$$= \underbrace{-A\,\varphi_r}_{\text{release to the lake}} + \underbrace{v_s\, A_s\, \pi P_L}_{\text{sediment.}} - \underbrace{r_s\, A_s\, P_{ads}}_{\text{transfer to permanent sediment}} \qquad (3)$$

where:

V_s	=	$A_s h_s$:
M	=	the total dry mass of solid sediment in the active layer $M = (1 - p)\,\rho_s\, V_s$.

P transfer to permanent sediment is assumed to take place at the rate determined by the sedimentation rate r_s which characterizes the lake and its catchment area, the solid mass M in the active sediment layer being assumed as constant.

Diffusion of dissolved P (PO_4) in pore water towards the permanent sediment is assumed to be negligible.

The concentration of dissolved P (PO_4) in pore water is considered as the "driving force" of P release. Hence, the most obvious assumption is that P release flux φ_r is proportional to the P concentration in pore water (P_{int}):

$$\varphi_r = v_r P_{int} \frac{V_{H20}}{V_s} = v_r P_{int} \, p = \frac{v_r p}{a} (P_{ads} - P_{ads,o}) , \tag{4}$$

the proportionality constant v_r with the dimensions of a velocity (m yr^{-1}), being empirically determined (note that introduction of a dependence of the release flux φ_r on the dissolved P concentration decrease at the sediment-water interface $P_{int} - (1 - \pi) P_L$ does not prevent the analytical solution of the balance equations; nevertheless this would be a unnecessary complication for our purposes since $P_{int} >> (1 - \pi) P_L$ during a large part of the evolution).

Then introducing (2) and (4), eq. (3) becomes:

(5)

Introducing (4) in (1) and putting $v_s\pi / h_s = k_{sed,s}$, the system of the linear differential equations which describe the evolution of the system lake + sediment is obtained:

$$\frac{dP_L}{dt} = - (k_{out} + k_{sed}) P_L + \frac{v_r p}{a \bar{z}} P_{ads} - \frac{v_r p}{a \bar{z}} P_{ads,o} + k_{in} P_{ext}$$

$$\frac{dP_{ads}}{dt} = \frac{k_{sed,s}}{\frac{p}{a} + (1-p) \rho_s} P_L - \frac{\frac{v_r p}{a} + r_{ads}}{h_s \left[\frac{p}{a} (1-p) \rho_s \right]} P_{ads} + \frac{\frac{v_r p}{a} P_{ads,o}}{h_s \left[\frac{p}{a} + (1-p) \rho_s \right]} \tag{6}$$

In general $(1\text{-}p)\rho_s >> \frac{p}{a}$: i.e. dissolved P in pore water is a negligible quantity in the sediment with respect to absorbed P; nevertheless its concentrations determines P release and magnifies all variations of P concentration in solid unsaturated sediment.

The solution of the system of eqs. (6) is given by the superposition of two exponentially decaying terms and an asymptotic term which gives the steady state concentration:

$$y\,(t) = y_1 \exp(-\alpha_1 t) + y_2 \exp(-\alpha_2 t) + y_\infty$$

where the coefficients y_1, y_2 are determined by the initial concentrations $y_1\,(0)$, $y_2\,(0)$.

The time constant which characterizes the slower transient term is:

$$\alpha_2 = \frac{1}{2}\left(a_{11} + a_{22} - \sqrt{(a_{11} - a_{22})^2 + 4\,a_{12}\,a_{21}}\right)$$

(the other time constant α_1 with the sign + before $\sqrt{\ }$),

where:

$$a_{11} = k_{out} + k_{sed}$$

$$a_{12} = \frac{v_r\,p}{a\,\bar{z}}$$

$$a_{21} = \frac{k_{sed,s}}{\frac{p}{a} + (1-p)\,\rho_s}$$

$$a_{22} = \frac{\frac{v_r\,p}{a} + r_s}{h_s\left[\frac{p}{a} + (1-p)\,\rho_s\right]}$$

The steady state concentrations, which can be considered as established after ~ $2\alpha_2^{-1}$, are given by:

$$P_{L\infty} = \frac{a_{22}\,a_{10} + a_{12}\,a_{20}}{a_{11}\,a_{22} - a_{12}\,a_{21}}$$

$$P_{ads,co} = \frac{a_{21}\,a_{10} + a_{11}\,a_{20}}{a_{11}\,a_{22} - a_{12}\,a_{21}}$$

where:

$$a_{10} = k_{in} P_{ext} - \frac{v_r p}{a \bar{z}} P_{ads,o}$$

$$a_{20} = \frac{\frac{v_r p}{a} . P_{ads,o}}{h_s \left[\frac{p}{a} + (1-p) \rho_s\right]}$$

The complete solutions of eqs. (6) are as follows:

$$P_L(t) = \frac{a_{21} \Delta P_L - (a_{22} - \alpha_2) \Delta P_{ads}}{a_{21} - (a_{22} - \alpha_2)(a_{11} - \alpha_1)/a_{12}} \exp(-\alpha_1 t) + (a_{22} - \alpha_2) .$$

$$\frac{a_{12} \Delta P_{ads} - (a_{11} - \alpha_1) \Delta P_L}{a_{12} a_{21} - (a_{22} - \alpha_2)(a_{11} - \alpha_1)} \exp(-\alpha_2 t) + P_{L\infty} ;$$

$$P_{ads}(t) = \frac{a_{21} \Delta P_L - (a_{22} - \alpha_2) \Delta P_{ads}}{\frac{a_{12} a_{21}}{a_{11} - \alpha_1} - (a_{22} - \alpha_2)} \exp(-\alpha_1 t) +$$

$$+ \frac{a_{12} \Delta P_{ads} - (a_{11} - \alpha_1) \Delta P_L}{a_{12} - (a_{22} - \alpha_2)(a_{11} - \alpha_1)/a_{21}} \exp(-\alpha_2 t) + P_{ads\infty} ;$$

were:

$\Delta P_L \quad = \quad P_L(0) - P_{L\infty}$;

$\Delta P_{ads} \quad = \quad P_{ads}(0) - P_{ads\infty}$.

After explicitation of the coefficients a_{ij} the following expressions are obtained:

$$\alpha_2 = \frac{1}{2}\left(k_{out} + k_{sed} + \frac{\frac{v_r\, p}{a} + r_s}{h_s\left[\frac{p}{a} + (1-p)\,\rho_s\right]} - \sqrt{\left(k_{out} + k_{sed} - \frac{\frac{v_r\, p}{a} + r_s}{h_s\left[\frac{p}{a} + (1-p)\,\rho_s\right]}\right)^2 + 4\,\frac{v_r\, p\, k_{sed,s}}{a\,\bar{z}\left[\frac{p}{a} + (1-p)\,\rho_s\right]}}\right) \qquad (7)$$

$$P_{L,\infty} = \frac{\left(1 + \frac{v_r\, p}{a\, r_s}\right) k_{in}\, P_{ext} - \frac{v_r\, p}{a\,\bar{z}}\, P_{ads,o}}{k_{out}\left(1 + \frac{v_r\, p}{a\, r_s}\right) + k_{sed}} \qquad (8)$$

$$P_{ads,\infty} = \frac{\bar{z}\, k_{sed}\, k_{in}\, P_{ext} + \frac{v_r\, p}{a}\, k_{out}\, P_{ads,o}}{\left(\frac{v_r\, p}{a} + r_s\right) k_{out} + r_s\, k_{sed}} \qquad (9)$$

In order to understand the physical meaning of these expressions, their limit has to be considered for release velocity v_r, tending to zero:

$$\lim_{v_r \to 0} \alpha_1 = k_{out} + k_{sed}$$

$$\lim_{v_r \to 0} \alpha_2 = r_s / [h_s\,(p/a + (1-p)\,\rho_s)]$$

$$\lim_{v_r \to 0} P_{L\infty} = k_{in} P_{ext} / (k_{out} + k_{sed})$$

$$\lim_{v_r \to 0} P_{ads\infty} = v_s \pi k_{in} P_{ext} / [r_s (k_{out} + k_{sed})].$$

I.e. lake and sediment evolutions are independent one of the other: the characteristic time of lake evolution is determined by outflow and sedimentation, that of sediment by the flux of sedimenting material and by the sediment structure, and is approximately coincident with the renewal time of the active sediment layer as defined above;

asymptotic phosphorus concentration in lake is the same as predicted by the mixed reactor approximation, while that in sediment is determined by the fraction of particulate phosphorus in the sedimentation flux.

With non vanishing release velocity the lake-sediment system evolution is dominated by a transient with characteristics time longer than $(k_{out}+k_{sed})^{-1}$, and even much longer, its higher limit being the renewal time of the active sediment layer.

The above described dynamics, as expressed by the second eq. (6), concern a sediment, which has not been saturated by phosphorus. In the latter case (P_{ads} = const.), the system can't be described any more by two equations, as the relation between P concentration in the pore water and P concentration in the solid sediment is no more valid. This corresponds to a highly anomalous situation, which is put in experimental evidence by very high dissolved P concentration in pore water, and even by a layer of unmineralized organic detritus at the sediment-water interface.

Transition from saturated to unsaturated sediment and assessment of the recovery time will be discussed in Part III with reference to the case of the hypereutrophic Lake Varese.

In the next step of complexity, a sediment model is associated to a lake model in which different phosphorus forms are taken into account separately. As a consequence, a sedimentary organic phosphorus compartment, which was neglected in the simplified model discussed above, is explicitely considered.

A possible extension of a lake model to sediments is briefly sketched at the end of ch.7.

Finally, a detailed description of chemical processes at the sediment-water interface, resulting either in phosphorus release or phosphorus precipitation, requires a much greater modelling effort, and should be conceived as structurally independent of a lake model.

In such a model the state variables include ionic forms of heavy metals (e.g. Fe), whose cycle is assumed to essentially affect release (or precipitation) of soluble (or insoluble) phosphates.

Sources and suggestions for further reading

1. M.W. Lorenzen, et al., A long term phosphorus model for lakes: application to Lake Washington.
   In "Modelling Biochemical Processes in Aquatic Ecosystems"(R.P. Canale, ed.), Ann Arbor Sc., pp 75-91, (1976)
2. G. Rossi and G. Premazzi, Delay in lake recovery caused by internal loading. Water Research, 25, 5, pp. 567-575, (1991)
3. P. Baccini, Phosphate Interactions at the Sediment-Water Interface.
   In "Chemical Processes in Lakes" (W. Stumm, ed.), J. Wiley and Sons, New York, pp. 189-205, (1984)

5. Modelling Thermal Stratification and Vertical Mixing

The ecological impact of lake-sediment interaction depends essentially on the influence of the released phosphorus on algal growth (secondary eutrophication). In deep lakes this influence becomes effective only when the released phosphorus has diffused from the bottom layers up to the euphotic zone. The diffusion process takes place along a water column, which exhibits a thermal (and density) stratification for much of the year, in a water body constantly dominated by turbulence phenomena.

As a consequence, a mixed reactor model as considered in previous chapters is certainly not adequate to describe lake phenomena having a short characteristic time with respect to the year (as for instance phosphorus cycles and algal growth), and at least a depth dependence of the state variables should be envisaged.

The simplest concepts underlying modelling the depth dependence are connected to thermal stratification of water and eddy diffusion of dissolved substances. For this reason, before one begins to construct a depth dependent lake model, it is useful to remind the elements of a physical theory of heat transfer in a turbulent medium.

Postponing to the next chapter a discussion on horizontal heterogeneities (of water temperature and dissolved matter concentration), water temperature T in a lake of section area A(z) at elevation z is assumed to depend only on z and t, and the *heat transfer equation* in the assumption of vanishing net heat flux from the walls, is written directly in the form:

$$A(z)\,\frac{\partial T}{\partial t}(z,t) = -\frac{\partial}{\partial z}\left\{A(z)\left[v(z,t)\,T(z,t) - \chi(z,t)\,\frac{\partial}{\partial z}T(z,t)\right]\right\} - \frac{1}{\rho c}\,\frac{\partial}{\partial z}\left[A(z)\,\varphi(z,t)\right] \qquad (1)$$

where ρ is the water density and c the water specific heat (both assumed as constant); v(z,t) a vertical water advection velocity; $\chi = \kappa/(\rho c)$ the vertical heat diffusion coefficient and κ the thermal conductivity of water; $\varphi(z,t)$ the solar radiation flux. The elevation z is counted positively upward, z = 0 being the water-sediment interface, $z = z_{max}$ the water-atmosphere interface. Eq.(1) represents the heat balance of the generic layer (z,z+dz) of the water body, taking into account advective, diffusive and solar radiation contributions.

The initial condition for Eq.(1) is supplied by a measured vertical temperature profile $T(z,t_o)$; boundary conditions are as follows: $T(0,t) = \text{const}(t)$, i.e. water temperature is assumed as constant at the lake bottom, which is acceptable for not too long a time interval; the condition at the surface is determined by the heat balance in a thin layer at the lake surface, i.e. by an equation of the type:

$$\beta\,\varphi\,(z_{max}, t) = -\,\rho c \chi(z_{max})\,\frac{\partial}{\partial z}\,T\,(z,t)\Big|_{z=z_{max}} - \varphi_L \tag{2}$$

where β is the long wavelength fraction of the incoming solar radiation ($\beta \cong 0.4$) and φ_L the total heat flux from the lake surface to the atmosphere (net long wave radiation + evaporation + conduction or "sensible" heat). Eq.(2) gives the value of

$$\frac{\partial}{\partial z}\,T\,(z,t)\Big|_{z=z_{max}}.$$

In Eq.(1) the source-sinks terms connected to inflow-outflow have been omitted for the sake of simplicity.

The advection term in the flux is expressed by means of a vertical advection velocity $v\,(z,t)$ determined by the water balance of each layer $(z,z+\Delta z)$:

$$\frac{\partial}{\partial z}\,[\,A\,(z)\,v\,(z,t)\,] = [\,q_{in}\,(z,t) - q_{out}\,(z,t)\,]/\Delta z\,, \tag{3}$$

where $q_{in}(z,t), q_{out}(z,t)$ are inflow-outflow rates. It is assumed that inflow concerns the water layer whose temperature is equal to the (measured) temperature of inflowing water at time t; outflow concerns the surface layer $(z_{max}, z_{max}-\Delta z)$, except for the case of artificial water withdrawal. The boundary condition for Eq.(3) at $z = z_{max}$ is supplied by the overall water balance of the lake.

The mathematical framework as presented above must be used to describe a phenomenology consisting of: (a) water volume heating due to penetrating (short wave-length) solar radiation; (b) heat exchange at the air-water interface concerning mainly long wave-length radiation; (c) vertical water mixing due to wind action and convective water movements, both associated with transformation of kinetic energy into potential energy and involving turbulence phenomena (turbulence generated by wind action and/or by instability of a local *shear* (vertical gradient of the horizontal water velocity du/dz)).

This phenomenology results in a characteristic thermal stratification pattern, with a water temperature profile (T vs z) exhibiting a maximum of the temperature vertical gradient dT/dz, whose magnitude and depth undergo an annual evolution (fig.1). Then thermal stratification of a lake refers in a strict sense to the existence of a stable temperature and density separation between surface layers (*epilimnion*) and deep

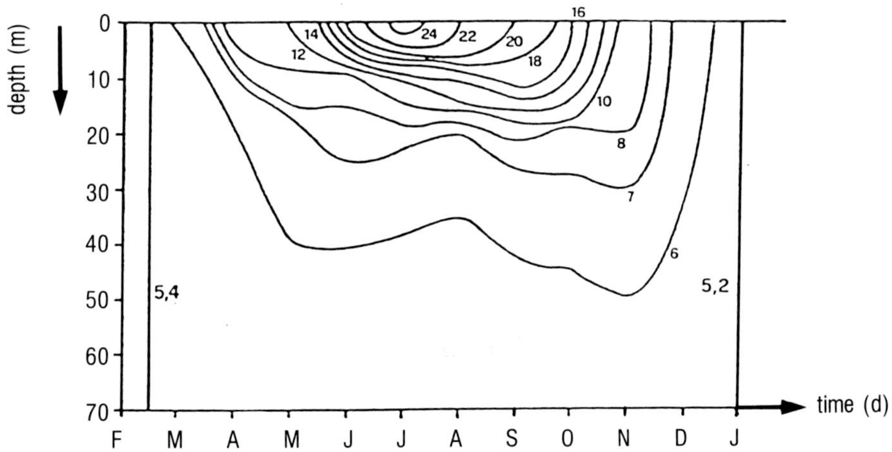

Fig. 1 Isotherms in Lake Lugano (Agno Bay) during an annual evolution. Winter omothermy, stratification and destratification are clearly exhibited.

layers (*hypolimnion*) while the interposed region of high gradients is called the *metalimnion.* Because the existence of such separation (by means of an ideal plane surface, the *thermocline)* depends on the equilibrium between wind action and buoyancy forces, the stability of the thermal stratification in a lake is expressed quantitatively by the *Richardson number,* defined as

$$Ri = \frac{\frac{g}{\rho}\left|\frac{\partial \rho}{\partial z}\right|}{\left(\frac{du}{dz}\right)^2} \tag{4}$$

where g is the gravity acceleration.

Eq.(4) means that stability of thermal stratification increases with increasing Ri. Moreover it is possible to establish a *stability criterion* Ri > ε, with a suitable value of ε, in the sense that the coexistence in a lake at time t of ρ(z),u(z) profiles satisfying the stability criterion in a given z-interval is a necessary (but not sufficient) condition for the existence of a stable laminar (not turbulent) flow in the z-interval concerned.

The key point for including the phenomenology described into Eq.(1) is clearly the function χ(z,t), which has been left undetermined until now, with the only implicit advice that its physical meaning is not strictly that of the thermometric conductivity of water (molecular diffusion coefficient of heat), approximately constant with respect to z and t (for this reason it is termed the *effective diffusion coefficient).*

In reality, vertical turbulent dispersion of heat is represented in the original thermal energy balance equation by the time average of the product of the fluctuations of water temperature and vertical water velocity, i.e. $< \delta T(t).\delta v(t)>_{\tau}$. The length of the time interval τ, on which the average is taken, is associated, according to the Kolmogorov theory, to the scale of lake water motions whose details are considered as unrelevant for the process we intend to describe.

The vertical flux $< \delta T(t).\delta v(t) >_{\tau}$ can be explicitly represented in various ways (*closure methods*), which are all more or less cumbersome.

The most usual method in ecological models is based on the *eddy diffusion coefficient* concept. By analogy with molecular diffusion, turbulent heat fluxes are given by the water temperature gradient, multiplied by a diffusion coefficient (eddy diffusion coefficient), a function of z and t, which is much greater than the molecular one:

$$< \delta T (z,t) . \delta v (z,t) >_{\tau} \equiv -\chi (z,t) \frac{\partial T}{\partial z} (z,t) \tag{5}$$

(note that time scale at r.h. side is not the same as at l.h.).

The concept of eddy (or effective) diffusion coefficient has rather shaky physical foundations (i.e. the analogy with molecular diffusion) and some disadvantages such as the entirely empirical determination of the dependence on z and t, and its non applicability to cases of vanishing gradients.

Nevertheless, its extensive use in ecological models suggests to give some indications on the determination of $\chi(z,t)$ on the basis of water temperature measurements (in the presence of vanishing temperature gradients, as in hypolimnion, one is forced to use other tracers, as for instance the excess activity of Rn - 222 in lakes with crystalline bedrocks).

The starting point for characterizing the vertical mixing properties of the metalimnion during summer stratification by means of an average eddy diffusion coefficient profile $< \chi(z,t) >_{\Delta t}$, related to a time interval Δt and to the required z-interval, is a formal solution of Eq.(1) with respect to $\chi(z,t)$. Assuming net heat flux equal to zero at the water-sediment interface, and neglecting the advective terms, this solution looks as follows:

$$\chi (z,t) = \frac{1}{A (z) \frac{\partial T}{\partial z} (z,t)} \left\{ \int_0^z A (\tilde{z}) \frac{\partial T}{\partial t} (\tilde{z},t) \, d\tilde{z} + \frac{A (z) \ \varphi (z,t)}{\rho c} \right\} \tag{6}$$

Clearly, Eq.(6) corresponds to use the heat balance of the water layer (0,z) to determine the diffusive heat flux across the surface A(z), then to determine $\chi(z,t)$, the temperature gradient $\partial T/\partial z$ at elevation z being known. The average value $< \chi(z,t) >_{\Delta t}$ is evaluated by introducing the averages $< \partial T/\partial z (z,t) >_{\Delta t}$, $< \partial T/\partial t (z,t) >_{\Delta t}$, $< T (z,t) >_{\Delta t}$,

$< \varphi (z,t) >_{\Delta t}$ into Eq.(6). These average values are related to a series of N water temperature profiles measured at times $t_1,...,t_N$, uniformly distributed in Δt.

The solar radiation flux profiles $\varphi(z,t)$ are calculated by means of an expression of the type:

$$\varphi (z,t) = \varphi (z_{max},t) \sum_{1}^{I} {}_i \, \varepsilon_i \, e^{-\alpha_i (z_{max},-z) \sec \psi (t)} \tag{7}$$

where ε_i,α_i are the energy fraction and extinction coefficient, respectively, related to the wave-length interval (i); $\psi(t) \equiv \arcsin(\sin \sigma(t)/n) \equiv$ refraction angle; n ≡ refraction index air-water (= 1.333); $\sigma \equiv$ incidence angle ≡ zenith angle.

The calculation of $\varphi(z_{max},t)$ is usually based on the measured solar radiation flux R_j incident at the lake surface during day j: the ratio $\varphi(z_{max},t)/R_j$ is expressed in terms of the daily average value of the cosine of the zenith angle (involving the photoperiod for day j) and of the albedo for the same day (see ch.8).

The choice of a "representative" time interval Δt and the averaging procedure, as for the quantity $\partial T/\partial t(z,t)$, require some care.

In fact, heat transfer in a lake is affected by atmospheric factors which exhibit episodic phenomena (e.g. the passage of a cold perturbation). Temperature patterns observed at a given depth (fig.2) are characterized by the superposition of oscillations of various frequencies on an "average" behaviour: this is particularly evident where vertical temperature gradients are large enough (say $\geq 1°C\ m^{-1}$). In general this indicates the existence of *internal gravity waves* (in particular standing internal waves or *internal seiches)*, which are the proximate source of turbulence at such depths.

Such temperature oscillations, which are due to wind perturbations, can affect the heat balance of the (0,z) layer, as is accounted for by Eq.(6), and can even lead to negative χ-values, since two-dimensional effects such as internal waves are not included in Eqs.(1) and (6).
Furthermore, the evaluation of the average water temperature increase per unit time at elevation z and during a time interval Δt ($< \partial / \partial t\ T (z,t)>_{\Delta t}$) requires a suitable linearization procedure on the measured data.
It is concluded that the effect of a series of wind perturbations on lake stratification can be correctly accounted for if the averaging procedure described above is applied to a sufficiently long time interval between two consecutive perturbations. The eddy diffusion coefficient profiles obtained are of the type shown in fig.3. The z-interval of the lowest χ-values corresponds to the thermocline depth for this period.

The ecologically relevant effect of thermocline lowering as a consequence of a succession of cold perturbations is the annexating to the mixed surface layers and to

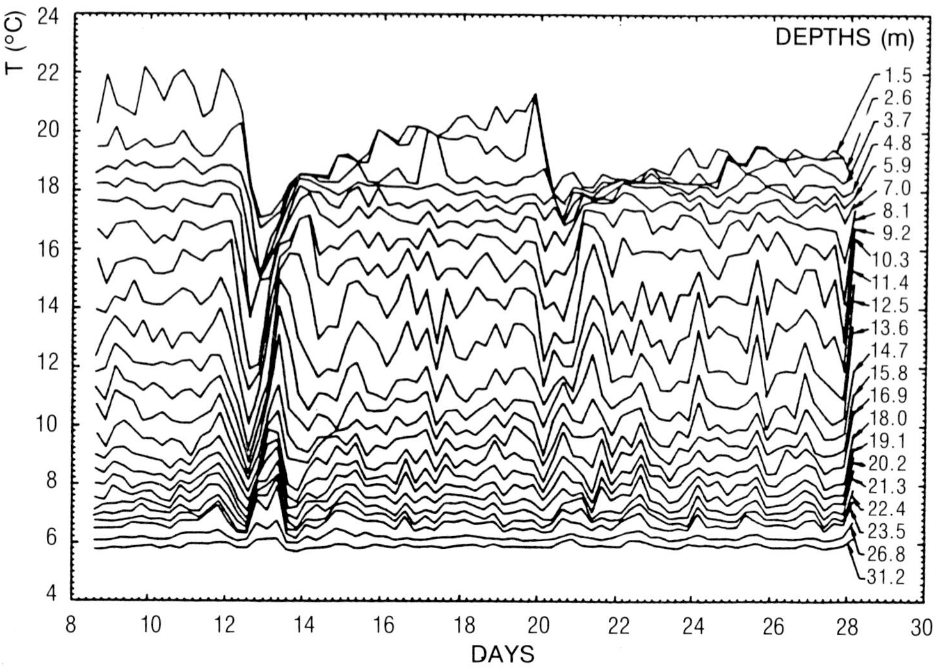

Fig. 2 Fluctuation of water temperature at various depths as measured and recorded in Lake Lugano in September 1978. The effect of a cold wind perturbation which interested the region on September 12th and 13th, is clearly exhibited.

the photic zone of hypolimnetic layers, which are richer in nutrients than epilimnetic ones in summer.

Calculations predicting the evolution of thermal stratification are based on a suitable *parameterization* of the z-dependence of the eddy diffusion coefficient. This was obtained with reference either to the square of the *Brunt-Väisälä frequency* $N^2 = |(g/\rho)(\partial\rho/\partial z)|$ which is the oscillation frequency of a water parcel vertically displaced from its equilibrium elevation in a stably stratified water body, or to the Richardson number Ri (Eq.(4)).

In the first case a relationship of the type

$$\chi(z) = a(N^2)^b \tag{9}$$

was adopted, and b was found close to the theoretically predicted value 0.5.

Remembering that in water $(1/\rho)(\partial\rho/\partial z) \cong -\beta dT/dz$ with

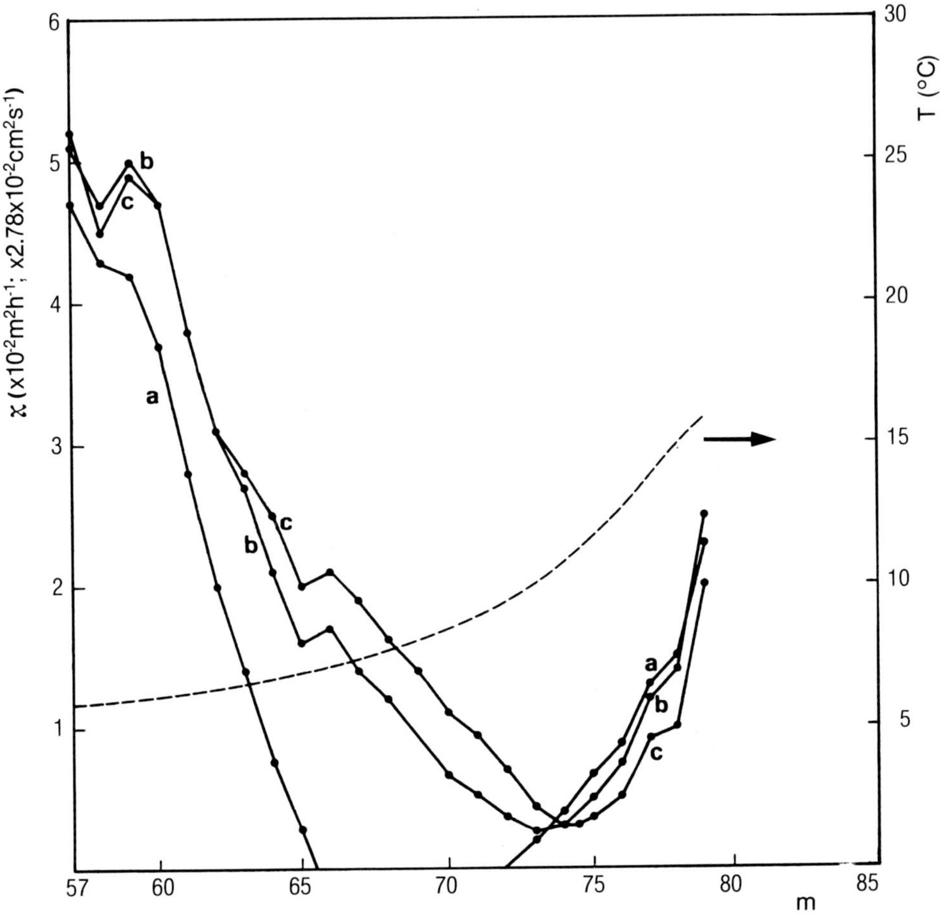

Fig. 3 Eddy diffusion coefficient as a function of elevation z, calculated on the basis of the measured temperatures of Fig.2. Curves a, b, c refer to averages on different time intervals. The time interval including the cold wind perturbation of the 12th and 13th September (curve a) originates negative values of the diffusion coefficient.

$\beta = 1.5\text{x}10^{-4\,°}\text{C}^{-1}$, Eq.(9) may be used to obtain a relationship between eddy diffusion coefficient and temperature gradient:

$$\chi\,(z,t)\left[\frac{\partial T}{\partial z}\,(z,t)\right]^{\frac{1}{2}} = \text{const}(t)\,, \qquad (10)$$

which is suited to recursive numerical procedures.

In the second case various relationships were proposed, such as:

$$\chi = \chi_o \exp(-.8\,Ri), \tag{11}$$

and:

$$\chi = \chi_o\,(1 + 3.33\,Ri)^{-3/2}, \tag{12}$$

where the value of χ for $Ri = 0$ ("neutral" stratification) was assumed as proportional to the mean shear $M = du/dz$ through the so-called *mixing length* ($\chi_o = \ell_m^2\, M$).

A totally empirical parameterization was also adopted in terms of two extreme values of χ connected by an exponential function of z:

$$\begin{aligned} \chi &= \chi_{max}\ \exp[-k\,(z_{max}-z)\,] & z \geq z_{th} \\ \chi &= \chi_{min} & z \leq z_{th} \end{aligned} \tag{13}$$

with $k = \log(\chi_{max}/\chi_{min}) / (z_{max} - z_{th})$.

Numerical solution of Eq.(1) with separately determined functions $v(z,t)$, $\chi(z,t)$ is currently obtained by means of implicit or explicit finite difference and finite elements methods.

The models include automatic modification of temperature profiles which present a maximum for $z < z_{max}$, to simulate correctly convective redistribution of heat in the surface layer. In addition to the "average" lake stratification behaviour, model predictions may concern the effect of the wind action during a time interval Δt on the thickness of the surface mixed layer. This requires the determination of the least elevation h which satisfies the relationship

$$g\int_h^{z_{max}} [\bar{\rho}-\rho\,(z)]\; z\; A\,(z)\; dz \leq E_c\,, \tag{14}$$

where $\rho(z)$, $\bar{\rho}$ are the density profile and the average density, respectively, in the interval (h, z_{max}) in the conditions existing before the wind perturbation, and E_c is the water kinetic energy produced by the wind action. For this quantity as a function of air density and wind velocity, various semi-empirical expressions have been proposed.

Sources and suggestions for further reading

1. J.C.J. Nihoul, Marine system analysis. Passive dispersion models.
   In "Modelling of Marine Systems" (J.C.J. Nihoul ed.), Elsevier, Amsterdam (1975)
2. A. Jassby and T. Powell, Vertical patterns of eddy diffusion during stratification in Castle Lake
   Limnol. Oceanogr., 20, 530-545 (1975)
3. G. Rossi, et al., On the evaluation of the vertical eddy diffusion coefficient in lakes by means of measured temperature profiles.
   In "Analysis of Ecological Systems: State-of-the-Art in Ecological Modelling" (W.K. Lauenroth, et al., eds.), Elsevier, Amsterdam, pp. 615-622 (1983)
4. R.E. Stauffer and G.F. Lee, The role of thermocline migration in regulating algal blooms.
   In "Modelling the Eutrophication Process" (E.J. Middlebrooks et al., eds.), pp. 73-82, Ann. Arbor Sc.
5. D.M. Imboden and S. Emerson, Natural radon and phosphorus as limnologic tracers: horizontal and vertical diffusion in Greifensee.
   Limnol. Oceanogr., 23, 77-90 (1978)

6. The General Equation of a Lake Eutrophication Model and the One-dimensional Approximation

The fundamental equation of a model describing the eutrophication process in a lake is a continuity equation for the constituent (i) distributed through the lake volume with a concentration $\rho_i(\vec{r}, t)$:

$$\frac{\partial \rho_i}{\partial t} + \mathrm{div}\,(\rho_i \vec{v}_i) = I_i(\rho_i, \rho_h, \ldots) + S_i \qquad (1)$$

where I_i,S_i summarize interaction and source-sink terms, respectively.

The velocity $\vec{v}_i$ of the constituent (i) particles is related to the velocity field of the lake water $\vec{v}(\vec{r}, t)$ by means of the identity

$$\vec{v}_i \equiv \vec{v} + (\vec{v}_i - \vec{v}) \equiv \vec{v} + \vec{w}_i$$

As a consequence the transport term div $(\rho_i \vec{v}_i)$ in Eq.(1) separates in an advection term div $(\rho_i \vec{v})$ describing the dragging of the constituent (i) by the water flow, and a dispersion term div $(\rho_i \vec{w}_i)$ describing the flux of constituent (i) inside the water body.

Modelling the eutrophication process concerns the time evolution of quantities, such as phytoplankton, nutrient, oxygen concentrations, for which time excursions of period smaller than one day can be considered inessential. Then in ρ_i, $\vec{v}$, $\vec{w}_i$ we separate the mean quantities from fluctuations:

$$\rho_i = \langle\rho_i\rangle + \delta\rho_i(t) \qquad \langle\rho_i\rangle = \frac{1}{\tau}\int_0^\tau \rho_i(t)\,dt \qquad \int_0^\tau \delta\rho_i(t)\,dt = 0$$

etc., and we choose $\tau \sim 1$ day.

The time integration $\int_{t_o}^{t_o+\tau} dt$ of Eq.(1) eliminates from the time derivative and other linear terms the fluctuating components $\delta\rho_i$, $\delta\vec{v}$, $\delta\vec{w}_i$.

Quadratic terms, such as $\rho_i\vec{v}$, $\rho_i\vec{w}_i$, give rise to the product of the averages, such as $\langle\rho_i\rangle\langle\vec{v}\rangle$, and to the average of the product of the fluctuations, such as

$\tau^{-1}\int_{t_o}^{t_o+\tau} \delta\rho_i\, \delta\vec{v}\, dt$. The sum of this term and the analogous one $\tau^{-1}\int_{t_o}^{t_o+\tau} \delta\rho_i \delta\vec{w}_i dt$, both under the div operator, represents the turbulent flux of the constituent (i), which is generally much larger than the flux due to the average flow. In order to express this flux in explicit form, one must face the same difficulty as for the thermal energy flux (ch.5). The elementary solution is to assume it as linearly dependent on the average concentration, in analogy with molecular diffusion, i.e. to put

$$\frac{1}{\tau}\,\mathrm{div}\int_{t_o}^{t_o+\tau} \delta\rho_i\,(\delta\vec{v} + \delta\vec{w}_i)\, dt \;\equiv\; -\,\mathrm{div}\,(\chi_i\ \mathrm{grad} <\rho_i>) \qquad (2)$$

where χ_i is the eddy (effective) diffusion coefficient for the constituent (i), whose value is assumed to coincide with that of the analogous coefficient in the heat transfer problem, once molecular diffusion has been neglected with respect to turbulent diffusion.

Taking into account the anisotropy of turbulent dispersion, the r.h. side of Eq.(2) takes the form:

$$\mathrm{div}\,(\chi_i\ \mathrm{grad} <\rho_i>) = \mathrm{div}_h\,(\chi_h\ \mathrm{grad}_h <\rho_i>) + \frac{\partial}{\partial z}\left(\chi_z \frac{\partial}{\partial z} <\rho_i>\right) \qquad (3)$$

where $\chi_h\ \mathrm{grad}_h \equiv \chi_x \frac{\partial}{\partial x}\vec{i} + \chi_y \frac{\partial}{\partial y}\vec{j}$ ($\vec{i}, \vec{j}$ are the unit vectors in the x, y directions).

Let all the symbols indicate time averages in the sense defined above, the fundamental equation of the model takes the form:

$$\frac{\partial \rho_i}{\partial t} + \underbrace{\mathrm{div}\,(\rho_i\vec{v})}_{\text{advection}} + \underbrace{\mathrm{div}\,(\rho_i\vec{w}_i)}_{\text{migration}} = \underbrace{I_i\,(\rho_i, \rho_h, \ldots)}_{\text{interaction}} + \underbrace{S_i}_{\text{source}} + \underbrace{\mathrm{div}\,(\chi_i\ \mathrm{grad}\ \rho_i)}_{\text{eddy diffusion}} \qquad (4)$$

$$(i = 1,\ldots,N)$$

where N is the number of the constituents considered.

Eqs.(4) are associated with initial and boundary conditions as for the concentrations ρ_i.

The most general treatment of the system of N partial differential equations (4) implies the knowledge of the mean water velocity field $\vec{v}(\vec{r}, t)$, i.e. the recourse to a (three-dimensional) hydro-dynamic model of the lake.

The migration term requires the determination of the mean velocity field $\vec{w_i}$ of particles of species i in their (typically settling) movement independent of water flow.

In the framework of an ecological model the complexity of the biologic interactions summarized by the term $I_i(\rho_i,\rho_h,...)$ suggests that Eqs.(4) should be simplified as much as possible.

Experimental investigation points out that vertical heterogeneities generally prevail in a lake over horizontal ones. It is then natural to think of an approximation in which model state variables are average values over a horizontal section, and the basic model equation is obtained from Eq.(4) by means of integration over a horizontal slice of the lake water body, whose thickness is made to tend to zero.

Moreover, in the framework of turbulence theories it is shown that the ratio between horizontal and vertical eddy diffusion flux vanishes as $10\,h/\ell$, where h is the thickness of the surface mixed layer, and ℓ the length scale of horizontal eddies. The fact that normally $\ell \gg h$ justifies neglect of the horizontal eddy diffusion flux in Eq.(3):

$$\mathrm{div}\,(\chi_i \,\mathrm{grad}\ \rho_i) \approx \mathrm{div}\,(\chi_{iz} \frac{\partial \rho_i}{\partial z} \vec{k}), \qquad (5)$$

where $\vec{k}$ is the unit vector in the z direction.

For these reasons, three and two-dimensional ecological models, requiring considerable computer work and being rather awkward to manage, have limited applications.

As has already been mentioned, the formal approach from Eq.(4) to the one-dimensional (z,t) model equation consists of integration of Eq.(4) over a horizontal slice of the lake water body, between two sections at elevations z_1 and $z_1 + \Delta z = z_2$. The boundary surface Σ of this slice includes the two horizontal sections S_1, S_2 plus the surface S_w of the strip of lake wall between them (fig.1). Green's formula is used in the form

$$\int_V \mathrm{div}\,\vec{v}\,dV = -\int_\Sigma \vec{v}\cdot\vec{n}\,dS \qquad (6)$$

where $\vec{n}$ is the unit vector normal to the boundary surface and directed inwards.

Then all the surface integrals at the r.h. side of Eq.(6) are subdivided into three pieces corresponding to surfaces S_1 (area $A(z_1)$), S_2 (area $A(z_2)$), and S_w, respectively. The quantities under the signs $\int_{S_1}$, $\int_{S_2}$ are replaced by average values on the corresponding horizontal surface everywhere; the quantities under the sign $\int_{S_w}$ are replaced by average values on the surface S_w, i.e. on the depth interval z_1, z_2.

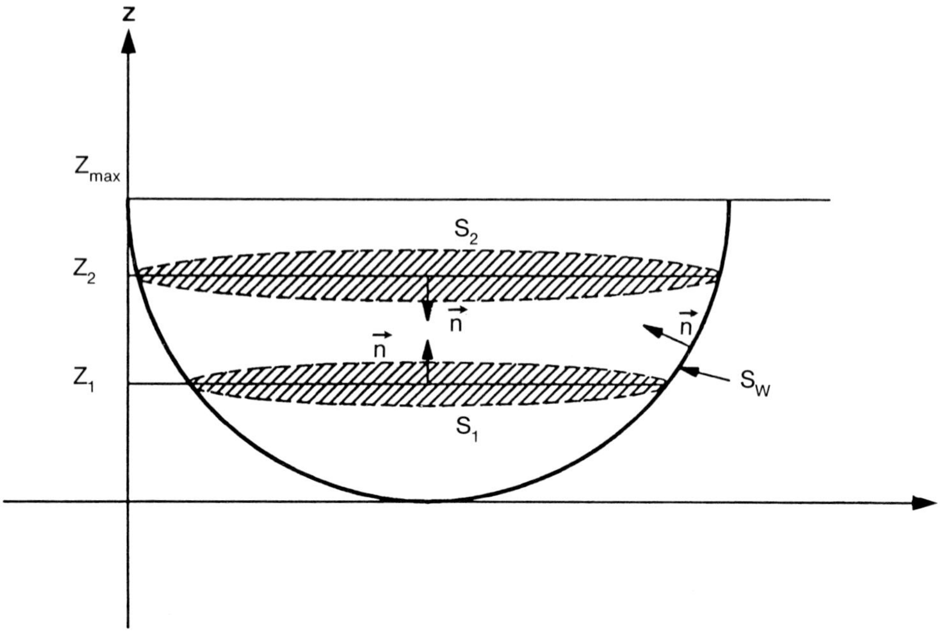

Fig.1 Scheme of the lake horizontal slice used for x,y-integration to derive the continuity equation of the (z,t) lake model.

In particular the term $\int_{S_w} \rho \vec{v} \cdot \vec{n}\, dS$ takes into account inflow-outflow and is transformed into a sum extended to tributaries (emissaries) of the inflow (outflow) rates q_{in} (q_{out}) times the respective concentrations, in the following way:

$$-\int_{S_w} \rho \vec{v} \cdot \vec{n}\, dS = -\sum_i \rho_i^{ext}\, q_{in}^{(i)} + \sum_j \rho\,(z_{max})\, q_{out}^{(j)} . \tag{7}$$

Taking into account Eq.(5), integration of the eddy diffusion term gives:

$$\int_V \mathrm{div}\left[\chi(z) \frac{\partial \rho}{\partial z} \vec{k}\right] dV = -\int_{\Sigma} \chi_z(z) \frac{\partial \rho}{\partial z} \vec{k} \cdot \vec{n}\, dS = -\chi_z(z_1) \frac{\partial \rho}{\partial z} A(z_1) +$$

$$+ \chi_z(z_2) \frac{\partial \rho}{\partial z} (z_2)\, A(z_2) + \varphi(z)\, [A(z_2) - A(z_1)] \tag{8}$$

where $A(z_2)\text{-}A(z_1) = \int_{S_w} \vec{n} \cdot \vec{k}\, dS$, and $\varphi(z)$ is the vertical component of the average flux of substance released from the lake walls between z_1 and z_2. It is assumed that the horizontal component of this flux has negligible effects on a pre-existing horizontally homogeneous distribution of that substance.

Then the continuity equation for a substance in a generic lake water layer $(z,z+\Delta z)$ is obtained, and making the Δz interval tend to zero the general equation of a one-dimensional (z,t) model is derived:

$$
\begin{aligned}
A(z)\frac{\partial \rho}{\partial t}(z,t) + \frac{\partial}{\partial z}\left[A(z)\, v_z(z,t)\, \rho(z,t)\right] + \frac{\partial}{\partial z}\left[A(z)\, w_z(z,t)\, \rho(z,t)\right] = \\
= \frac{\partial}{\partial z}\left[A(z)\, \chi_z(z,t)\frac{\partial \rho}{\partial z}(z,t)\right] + \varphi(z,t)\frac{dA}{dz} + \\
+ (I+S)\, A(z) + \left[\rho_{ext}(t)\, q_{in}(t) - \rho(z,t)\, q_{out}(t)\right]\cdot\frac{1}{\Delta z}
\end{aligned}
\qquad (9)
$$

where $\rho(z,t)$ is to be understood as a horizontally averaged value at elevation z and the last term is present only at $z = z_{max}$.

The application of a model to a particular lake requires, after determination of the functions $A(z), v(z,t), w(z,t), \chi_z(z,t)$ for that lake, the choice of N state variables, i.e. concentrations $\rho_i(z,t)$, which are assumed to give a realistic picture of the trophic evolution of the lake: e.g. one or many "groups" of algal species, the observed chemical forms of the limiting nutrient, dissolved oxygen, etc.

The interaction term $I(\rho_i,\rho_h,...)$ for each of these variables with respect to the others is the result of modelling bio-chemical processes, which are relevant to the evolution of the ecosystem, such as algal growth, phosphorus uptake, organic phosphorus mineralization, etc. Phosphorus release from sediments is accounted for by a function $\varphi(z,t)$, which is the result of modelling lake-sediment interaction as discussed in ch.4, as well as of field and laboratory investigation.

Detailed information concerning these *submodels* is postponed to the next chapter, where the box-approximation to the (z,t) model is presented and discussed.

Systems of N differential equations as (9) are numerically treated by essentially the same methods as those used for the solution of the heat transfer problem, with the additional difficulties associated with the non linearities arising from the interaction terms $I(\rho_i,\rho_h,...)$.

One-dimensional (z,t) models have been quite extensively applied to eutrophicated lakes for both interpretative and predictive purposes as for instance to Lake Washington (Chen and Orlob) and to the Swiss lakes Alpnachersee and Greifensee (Imboden and Gaechter).

As an intermediate approximation between (r,t) and box models, it is worth mentioning the simulation of Lake Kootenay biochenosis (R.A. Parker), in terms of differential equations referred to the surface mixed layer, in which horizontal and vertical transport was accounted for by introducing source terms $v \frac{\partial}{\partial x} C(x,z,t)$, $m \frac{\partial}{\partial z} C(x,z,t)$ in the generic (x,z) point of the layer. The constants (with dimensions of velocity) v,m were experimentally determined and the gradients were estimated assuming linear dependence on the distance from the next point in the lake, where concentrations had known values.

Sources and suggestions for further reading

1. J.C.J. Nihoul, Marine System Analysis. Passive dispersion models.
   In "Modelling of Marine Systems" (J.C.J. Nihoul, ed.), Elsevier, Amsterdam (1975).
2. D.M. Imboden and R. Gächter, A dynamic model for trophic state prediction.
   Ecol. Model., 4, 77-98 (1978).
3. R.A. Parker, Some problems associated with computer simulation of an ecological system.
   In "Mathematical Theory of the Dynamics of the Biological Populations" Oxford University, pp. 269-288 (1973)

7. The Box Approximation

Eddy diffusion coefficient profiles $\chi(z)$ of the type presented in fig.3 of ch.5 synthesize a situation in which external actions (solar radiation and wind) and convective water movements due to surface cooling during the night, have created a surface mixed layer (epilimnion) which is clearly separated from the underlying stagnant layers (hypolimnion). The region between, characterized by high temperature and density gradients (metalimnion), is usually idealized as a horizontal surface (thermocline).

This is the physical basis of a common simplification of the one-dimensional (z,t) model discussed in the previous chapter, the so-called zero-dimension or box-model. Its fundamental assumption is the uniform (horizontal and vertical) distribution of the substances considered in each region (i.e. epilimnion and hypolimnion). Because of this assumption the calculational procedure is considerably simplified, which allows both a more detailed description of the relevant bio-chemical processes and a more direct interpretation of the results.
On the other hand, space uniformity in a given lake region can be considered as a realistic assumption, provided that the following conditions are statisfied:

(1) the water mean residence time in the region volume is much greater than the characteristic time of the physical (diffusion) and biochemical (nutrient cycle) processes occurring inside it (in particular: inflowing water, with its content of "external" substances, is mixed with residual water within a time, which can be considered as short with respect to the water mean residence time);

2) the characteristic time of (horizontal and vertical) effective diffusion is short with respect to that of the considered biological and chemical processes (this condition is poorly respected, for instance, in the case of phytoplankton growth especially as depth dependent).

The equations of a vertical box model (two for each state variable except for living phytoplankton biomass, which is negligible in the hypolimnion) are formally derived by means of z-integration of Eqs.(6.9) over the epilimnetic (z_{th},z_{max}) and hypolimnetic (0,z_{th}) z-intervals. The thermocline elevation z_{th} is obviously time-dependent, but in order to preserve the advantages of simplicity of the model, it is usual to adopt a fixed (average summer) value of z_{th}, and to impose on the eddy diffusion coefficient $\chi(z_{th},t)$ a time evolution which ensures total vertical mixing and vertical homogeneity of dissolved substances over the whole water column in winter. Alternatively, it is possible to take separately account of thermocline lowering effects on epilimnetic and hypolimnetic concentrations, by introducing terms such as $- dV_1/dt\ C_1$, $- dV_2/dt\ C_2$

$= dV_1/dt\ C_2$ in the r.h. sides of the balance equations,once the derivative $\frac{d}{dt} z_{th}$ has been properly evaluated.

In practice the model equations are directly written as balance equations for the various eco-system constituents in the epilimnion and hypolimnion, respectively, following the usual scheme: total mass variation per unit time = (volume + surface) source intensity - (volume + surface) sink intensity.

As for the various terms of the mass balance the following suggestions must be taken into account.

Eddy diffusive exchange between the epi- and hypolimnion is expressed in terms of the average epi- and hypolimnetic concentrations ρ_{i1}, ρ_{i2}, as follows:

$$- A (z_{th}) < \chi (z_{th} , t) >_{\Delta t} \frac{\partial \rho_i}{\partial z} \Big|_{z_{th}} = A (z_{th})\, m_{th} (\rho_{2i} - \rho_{1i}) \,. \qquad (1)$$

The *mixing coefficient at the thermocline depth* m_{th} (m d^{-1}) is given by the ratio χ/ℓ, where the distance ℓ (a sort of mixing length) has the obvious physical interpretation of metalimnion thickness.

Settling of particulate substances is expressed by the migration flux across the thermocline, i.e. by the product of the related concentration and a *settling velocity* v_s. This depends on the type of particle and on the water density (temperature), i.e. it should take different values in the epi- and hypolimnion (it is worth mentioning that dispersion of suspended particles can be associated also with vertical mixing by turbulence).

Sedimentation flux in epilimnion concerns the horizontal projection of epilimnetic walls, i.e. a surface of area $A(z_{max}) - A(z_{th})$; in the hypolimnion it concerns the horizontal projection of the whole hypolimnetic cuvette, i.e. the surface of area $A(z_{th})$.

Interaction terms in each equation represent the rate of production or destruction of a constituent due to biological or chemical processes. They must be given appropriate expressions in terms of concentration of interacting ecosystem constituents and of environmental parameters (such as water temperature and light intensity). Their specification is consequent to the choice of the constituents (state variables) and of the relevant interactions, as well as to the formulation of the laws of these interactions.

In the following, reference is made to a eutrophication model aiming at the description of the phosphorus cycle in its annual evolution, i.e. to the extension of the mixed reactor model of ch.2 to take account of biomass evolution as a sink for inorganic dissolved phosphorus and a source for organic particulate phosphorus, as well as of the heterogeneities due to thermal stratification and surface sources or sinks.

The ecosystem constituents whose mass balance is accounted for in such a model are:

1) phytoplankton biomass (one or two groups of algal species) (B);

2) intracellular (algal) phosphorus (N_o);

3) dissolved inorganic phosphorus (orthophosphate P-PO_4) (N_1);

4) particulate organic phosphorus (N_2);

5) dissolved oxygen (Ω).

The scheme of the interactions between these constituents is presented in fig.1, and the corresponding balance equations (1-8)are given in tab. I, where for the sake of simplicity only the essential aspects have been considered.

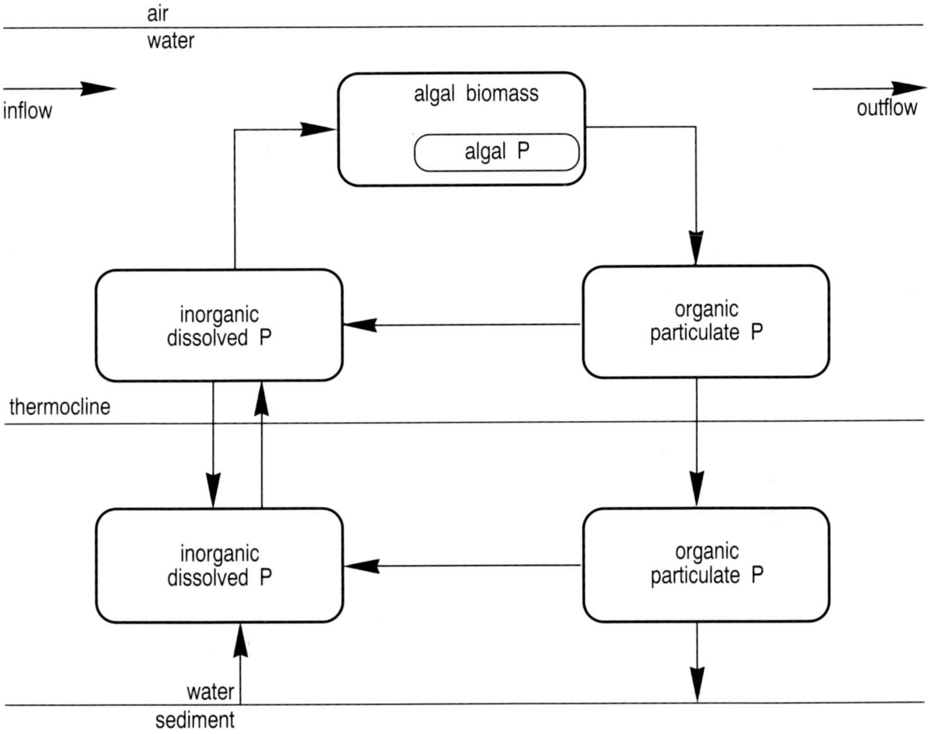

Fig. 1 State variables and interactions of a two vertical box eutrophication model (dissolved oxygen and related interactions have been omitted for clarity reasons).

Tab. I Equations of a Two Vertical Box Model

$$V_1 d/dt\, B = (\mu - r - s)\, BV_1 - q_{out}\, B \quad (1)$$

$$d/dt\, N_o = u - \mu N_o \quad (2)$$

$$V_1 d/dt\, N_{11} = q_{in} N_1^{\,ext} - q_{out} N_{11} - uBV_1 + k_{min,1} N_{21} V_1 + + A(z_{th})\, m_{th}\, (N_{12} - N_{11}) + \left[A(z_{max}) - A(z_{th})\right] \varphi_1 \quad (3)$$

$$V_1 d/dt\, N_{21} = q_{in} N_2^{\,ext} - q_{out} N_{21} + sN_o BV_1 - v_s A(z_{max})\, N_{21} - k_{min,1} N_{21} V_1 \quad (4)$$

$$V_1 d/dt\, \Omega_1 = A(z_{max})\, k_{aw}\, (\Omega_{sat} - \Omega_1) + k_\Omega (\mu - r)\, N_o BV_1 + A(z_{th})\, m_{th}\, (\Omega_2 - \Omega_1) - k_\Omega k_{min,1} N_{21} V_1 - \left[A(z_{max}) - A(z_{th})\right] \varphi_{\Omega 1} - q_{out}\, \Omega_1 - \beta_\Omega \quad (5)$$

$$V_2 d/dt\, N_{12} = k_{min,2} N_{22} V_2 + A(z_{th})\, m_{th}\, (N_{11} - N_{12}) + A(z_{th})\, \varphi_2 \quad (6)$$

$$V_2 d/dt\, N_{22} = -k_{min,2} N_{22} V_2 + A(z_{th})\, v_s N_{21} - A(z_{th})\, v_s N_{22} \quad (7)$$

$$V_2 d/dt\, \Omega_2 = A(z_{th})\, m_{th}\, (\Omega_1 - \Omega_2) - k_\Omega k_{min,2} N_{22} V_2 - A(z_{th})\, \varphi_{\Omega 2} \quad (8)$$

Postponing the discussion of the interaction laws to the next chapter, the fundamental quantities involved in the coefficients of the model equations are defined below.

The *phytoplankton growth coefficient* μ (daily production of algal biomass relative to existing algal biomass, dimensions d^{-1}) is connected to the efficiency of the photosynthetic process, as it occurs in the average physico-chemical conditions of the epilimnion at day t. Although a distinction between epilimnion and *photic zone* is often considered as unnecessary in box-models of lake eutrophication, parameters affecting phytoplankton growth are better referred to the latter, as being the region where the most stringent limitation to photosynthetic activity (that connected to light intensity) is satisfied ($V_1 = V_{ph}$ in Eq.(1) of Table I).

The volume V_{ph} of this zone varies with time in a way which can be assumed as independent of thermocline depth variation. Once photic depth has been defined and calculated, a term $-d/dt\, V_{ph}.B$ can be introduced into the r.h.s. of Eq.(1). As defined above, μ is a "gross" coefficient, in which biomass losses due to metabolic activities (the *respiration rate* r) and *mortality* (s) are not included.

The fundamental quantity determining the interaction between algal biomass and nutrient is the *phosphorus uptake velocity* u (referred to existing algal biomass, hence with dimensions d^{-1}). Regeneration of nutrient as a result of bacterial activity on organic detritus is expressed by means of the *mineralization rate* k_{min}, as the fraction of organic particulate phosphorus transformed daily into inorganic dissolved (reactive) phosphorus.

Release from sediment is accounted for by orthophosphate *phosphorus fluxes* φ_1,φ_2 g m^{-2} d^{-1}, related to epilimnetic and hypolimnetic sediment, respectively.

Finally, in the mass balances for dissolved oxygen coefficients appear such as k_{aw} (*air-water mixing coefficient*, m d^{-1}) and k_Ω (ratio oxygen produced/phosphorus consumed in the photo-synthetic reaction), as well as the *oxygen demand of inflowing water* β_Ω, and the outgoing oxygen flux $\varphi_\Omega(t)$ which represents the *oxygen demand by sediment*, i.e. daily oxygen consumption per unit sediment area at the water-sediment interface, and can be assumed as proportional to the phosphorus release flux $\varphi(t)$.

As announced at the end of ch.4, balance equations for the phosphorus forms associated to the active sediment can be added to the equations of tab. I. The purpose of such an *extended box model* are the following: 1) to predict the evolution of the potentially available phosphorus content of sediment; 2) to connect release fluxes with the sedimentary dissolved phosphorus concentration.

Referring us to the concepts discussed in ch.4 and taking into account that phosphorus dissolved in interstitial water is fed by organic phosphorus mineralization (at the rate $k_{min,s}$), balance equations for organic phosphorus (i.e. phosphorus associated to sedimented organic detritus) and total inorganic phosphorus concentration in the homogeneously mixed active sediment layer are as follows (see the scheme of fig.2):

$$M\, d/dt\, P_{org} = -\, k_{min,s} M P_{org} + f v_s A P_{part}$$

$$d/dt\, (pV_s P_{int} + MP_{ads}) \equiv (pV_s + aM) d/dt\, P_{int} = k_{min,s} M P_{org} - $$

$$- v_{rel} A p P_{int} + (1-f) v_s A P_{part} - r_{sed} A P_{ads,0} \qquad (3)$$

where:

M	=	total solid mass in active sediment volume V_s with porosity p;
A	=	area of sediment surface;
$v_s P_{part}$	=	phosphorus sedimentation flux;
f	=	organic fraction in sedimentation flux;
r_{sed}	=	sedimentation rate;
$v_{rel} P_{int}$	=	release flux;
$P_{ads,0}$	=	phosphorus concentration at the bottom of active sediment layer.
a	=	distribution coefficient ($P_{ads} = aP_{int} + P_{ads,o}$)

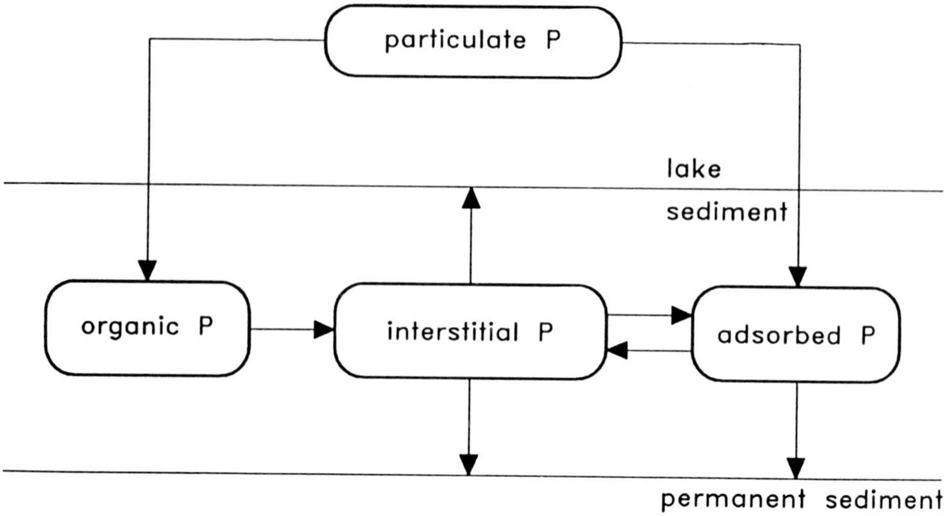

Fig. 2 State variables and interactions of a simplified sediment model.

Eqs.(3) can represent a valuable interpretative and predictive scheme for lake-sediment interaction, after a suitable procedure has been established for the determination of parameters f, v_{rel}, $k_{min,s}$.

In spite of the inherent drastic simplifications, box models have been widely used, especially as a basis for the discussion of process interdependence.

Vertical box models with three boxes have been envisaged to properly describe hypolimnetic vertical heterogeneities, as for instance those determined by intense nutrient release from bottom sediment.

Finally, mixed horizontal-vertical box models can be constructed to discuss more complex heterogeneities, such as occur in lakes with shallow bays (fig.3). Clearly each additional interface introduced raises the problem of determining the respective mixing coefficient.

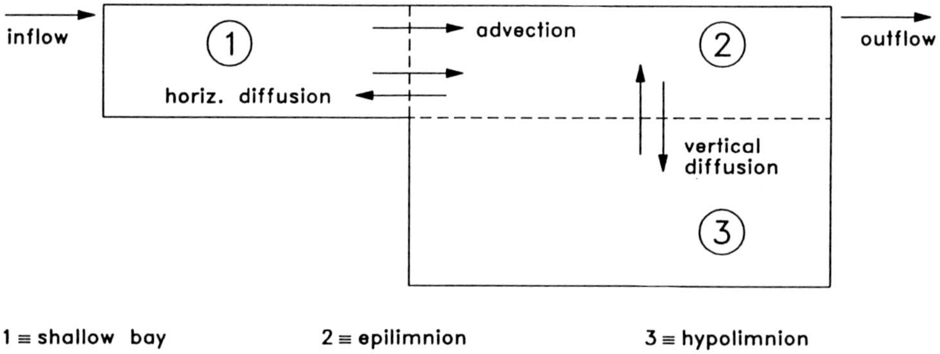

Fig. 3 Scheme of a mixed horizontal-vertical box model.

Sources and suggestions for further reading

1. D.M. Imboden and R.P. Schwarzenbach, Spatial and temporal distribution of chemical substances in lakes: modelling concepts.
   In "Chemical Processes in Lakes" (W. Stumm, ed.), J. Wiley and Sons, New York, pp. 1-30 (1984).
2. D.M. Imboden, Phosphorus models of lake eutrophication.
   Limnol. Oceanogr., 19, 297-304, (1974)
3. G. Rossi, et al., Correlation of a lake eutrophication model to field experiments, Ecol. Model., 34, pp 167-189 (1986)

8. Modelling Algal Growth and Bio-Chemical Interactions

The dependence of the algal growth coefficient μ on the *limiting factors* of the photosynthetic process: water temperature T, light intensity I and intracellular phosphorus concentration N_o, is expressed as a product of functions of these quantities:

$$\mu \sim f_1 (T) . f_2 (I) . f_3 (N_o), \qquad (1)$$

which means that the limiting factors act independently on the efficiency of the photosynthetic process.

Eq.(1) is equivalently presented in the form

$$\mu = \mu_{max} (T) . f (I) . g (N_o), \qquad (2)$$

where light intensity and intracellular nutrient concentration appear as limiting factors for a process whose maximum efficiency is dependent on the water temperature.

Adopting this approach, the *temperature dependence* of the maximum growth rate is determined on the basis of experiments on single algal species cultures in optimal light and nutritional conditions. The experiments allow identification of an optimal growth temperature, which is characteristic of each algal species. In natural conditions the evolution of a phytoplankton population in a water body of increasing (spring, summer) and decreasing (autumn, winter) temperature, can be considered as a succession of algal species blooms, whose optimal growth temperature coincides with the epilimnetic water temperature at that time.

In a limited temperature interval (e.g. $T \leq 25°C$) this dependence of optimal growth rate on temperature can be approximately represented by means of an exponential function:

$$\mu_{max} = a \exp (bT). \qquad (3)$$

Hence modelling the seasonal evolution of phytoplankton population in a lake starts from the determination of the constants (a,b) suited to the trophic stage of that lake. The following values were suggested: a = 0.59; b = 0.0633, as giving an upper physiological limit to phytoplankton growth, experimentally determined in cases in which neither light intensity nor nutrient concentration were considered as limiting (fig. 1).

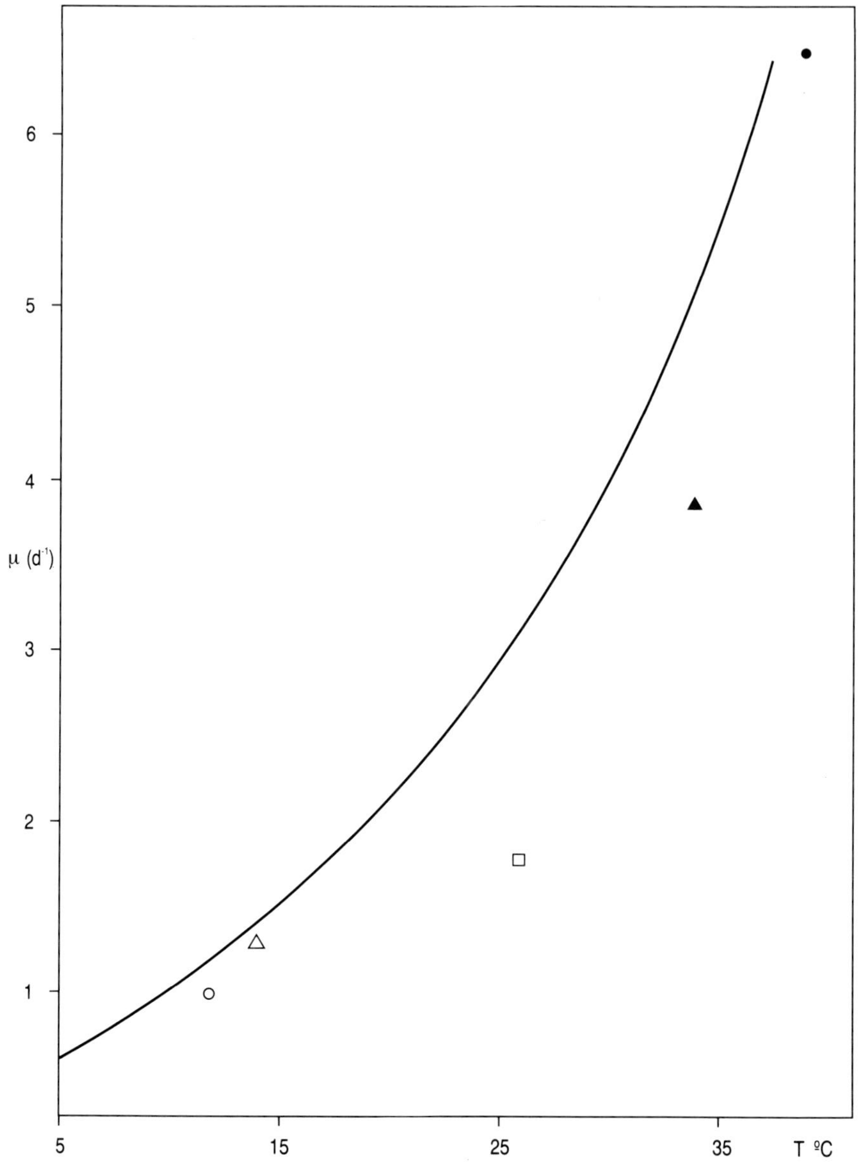

Fig. 1 Temperature dependence of the algal growth coefficient in optimal illumination and nutritional conditions (μ_{max} vs T).
Experimental points refer to 5 algal species (○ Detonula confervacea, △ Skeletonema costatum, □ Ditylum brightwellii, ▲ Dunaliella tertiolecta, ● Chlorella pyrenoidosa).
The curve (y = 0.50 exp (0.0633 x) was proposed by Eppley to represent the temperature dependence of specific maxima.

The temperature dependence of Eq.(3) can be equivalently put in the form

$$\mu_{max} = \mu_{max}(T = T_o)\,\Theta^{T-T_o}, \qquad (4)$$

T_o being usually taken equal to 20°C, and Θ being chosen on the basis of a measured temperature coefficient $\frac{1}{\mu}\frac{d\mu}{dT}$.

The determination of a function f(I) which can express the *light intensity dependence* of algal growth coefficient in the photic zone of a lake consists of the following steps: 1) calculation of light intensity in the highest water layer ($z = z_{max}$) as a function of time (day) in the annual evolution; 2) evaluation of light extinction, i.e. calculation of light intensity at elevation z within the lake, as a function of the concentration of phytoplankton biomass present at time t (shadow effect of algae); 3) adoption of a function $\theta[I(z,t)]$ to express the dependence of the algal growth rate μ on light intensity at elevation z and time t; 4) calculation of the average of $\theta[I(z,t)]$ in the photic zone. Hence the function f(I) is obtained:

$$f(I,t) = \frac{1}{z_{max} - z_{ph}} \int_{z_{ph}}^{z_{max}} \theta[\,I(z,t)\,]\,dz \qquad (5)$$

As for the calculation of the average solar radiation flux incident into the lake surface at day t (i.e. at zenith angle $\sigma = \lambda - \lambda_z$, where λ is the latitude and λ_z the solar declination at day t), an expression of the following type is currently adopted:

$$I(z_{max},t) = I_o(1-\rho)(1-0.54c^2)\exp\left[\frac{-(\alpha h)_{atm}}{<\cos\sigma>_t}\right] . <\cos\sigma>_t \qquad (6)$$

where:

ρ	=	the albedo of the lake surface (a function of σ, ψ)
c	=	the fraction of cloudiness ($0 \le c \le 1$);
$<\cos\sigma>_t$	=	$\sin\lambda \sin\lambda_z + 2/\gamma \cos\lambda \cos\lambda_z \sin\gamma/2$
$12\gamma/\pi$	=	the photoperiod (hours)at day t;
$\psi(t)$	=	arcsin [sin $\sigma(t)/n$] where n is the air-water index of refraction;
$(\alpha h)_{atm}$	=	the "optical height" of the atmosphere;
I_o	=	the solar radiation flux incident at the top of the atmosphere in the wavelength interval relevant to photosynthesis (chlorophyll absorption: 400-700 nm);
I_o	=	$(q_o/r^2)\ <\cos\sigma>_t$, where
q_o	=	solar constant ~ 20 kcal $m^{-2}min^{-1}$;
r	=	ratio actual earth-sun distance/average distance ($0.9669 < r^2 < 1.0037$).

Light extinction evaluation is based on the well-known experimental law of Beer and Lambert:

$$I(z,t) = I(z_{max},t)\exp[-k(z_{max}-z)] \tag{7}$$

k being the extinction coefficient (m^{-1}) assumed as an intrinsic property of the water column and as linearly dependent on the phytoplankton biomass concentration B, via the relationship

$$k = k_o + k_1 B(t)\,. \tag{8}$$

Assuming all the incident solar radiation as direct, the average length of the light path to reach depths $z_{max}-z$ at day t is given by

$$d = (z_{max}-z)/ < \cos\psi >_t, \tag{9}$$

ψ being the refraction angle corresponding to an incidence angle determined by zenith angle $\lambda-\lambda_z$ and hour angle ω, the average on ω being extended to the photoperiod of day t.

A commonly accepted form for the light dependent factor in the algal growth coefficient at z and t was proposed by Steele:

$$\theta(z,t) = \frac{I(z,t)}{I_{opt}}\cdot\exp\left[1-\frac{I(z,t)}{I_{opt}}\right], \tag{10}$$

I_{opt} being the light intensity at which photosynthetic efficiency is maximum at a given temperature and in a given nutritional environment (fig.2; Eq.(10) accounts for a inhibitory light effect). After introduction into Eq.(10) of the daily time average I(z,t), obtained by using Eqs.(6), (7) and (9), the z-integral [Eq.(5)] is easily evaluated:

$$f\left[I(z_{max})\right] = \frac{<\cos\psi>_t e}{k(z_{max}-z_{ph})}\left\{\exp\left[-\frac{I(z_{max})}{I_{opt}}e^{-\frac{k(z_{max}-z_{ph})}{<\cos\psi>_t}}\right]-\exp\left[-\frac{I(z_{max})}{I_{opt}}\right]\right\} \tag{11}$$

The function f [I (z_{max})] is found to be not very different from a simple hyperbolic dependence of the Michaelis-Menten type, i.e.

$$f(I) = I/(k_I+I) \tag{12}$$

with a suitable half-saturation constant k_I.

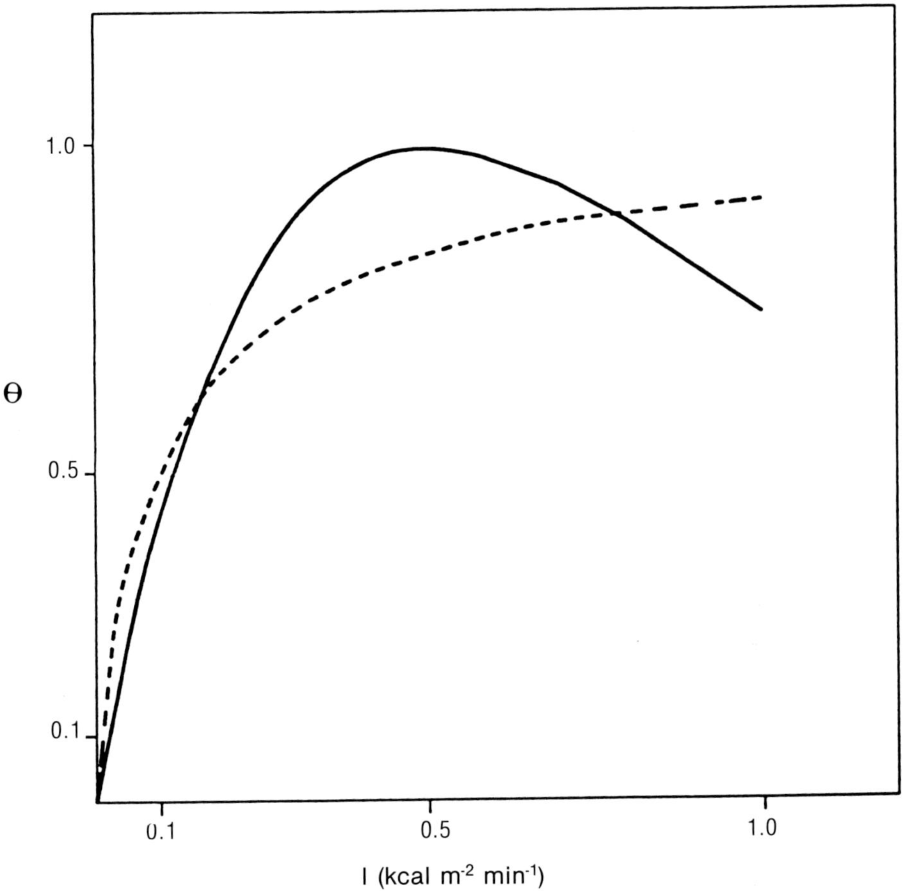

Fig. 2 Light intensity dependence of photosynthetic efficiency. Continuous curve represents the form proposed by Steele (Eq.(10)) with I_{opt} = 0.5 kcal m^{-2} min^{-1}. Dashed curve represents the hyperbolic dependence (Eq.(12)) with k_I = 0.1 kcal m^{-2} min^{-1}.

In fact, a function such as (12) with I equal to the average light intensity in the photic zone at day t, has been considered to represent an acceptable approximation for highly eutrophic lakes, where light inhibition hardly ever occurs except for a rather thin surface layer.

The dependence of the algal growth coefficient on the *intracellular phosphorus concentration* is expressed by means of a hyperbolic function of N_o-$N_{o,min}$ as follows:

$$g\,(N_o) = \frac{N_o - N_{o,min}}{k_{N_o} + (N_o - N_{o,min})} \qquad (13)$$

where the specific constant $N_{o,min}$ represents the intracellular concentration of phosphorus not available to the photosynthetic process.

Eq.(13) is often used in the simplified form with $k_{N_o} = N_{o,min}$, i.e.:

$$g(N_o) = 1 - N_{o,min}/N_o \; . \tag{14}$$

A direct connection between algal growth rate and external dissolved inorganic phosphorus concentration N_1 is contradicted by experimental results on algal growth in a nutrient depleted environment. Nevertheless a simple Michaelis-Menten factor, i.e.

$$g = (N_1) = N_1/(k_{N_1} + N_1) \tag{15}$$

is considered to represent an acceptable approximation in cases in which nutrient depletion never occurs.

External (dissolved inorganic) phosphorus concentration affects algal growth through the *phosphorus uptake process,* as it is described by the phosphorus uptake velocity u. This is determined by the external dissolved phosphorus concentration, and limited by phosphorus accumulated inside the cell. The dependence on external phosphorus concentration is commonly expressed by means of a Michaelis-Menten hyperbolic function, while the maximum uptake rate is determined by the intracellular phosphorus concentration through a linear feed-back factor:

$$u = u_{max} \frac{N_{o,max} - N_o}{N_{o,max} - N_{o,min}} \cdot \frac{N_1}{k_{N_1} + N_1} \tag{16}$$

Nutrient uptake velocities as dependent on external and internal nutrient concentrations are presented in fig.3.

The factor dependent on external nutrient concentration in Eq.(16) can be obtained as a limit case of a more elaborate algal cell model, which considers two intracellular phosphorus compartments (dissolved and “storage” phosphorus). Uptake is determined by an equilibrium reaction between phosphate (external and internal dissolved) and a membrane carrier molecule.

The resulting expression for phosphorus uptake velocity is

$$u = u_{max} \left(\frac{1}{1 + k_a N_{o,diss}} - \frac{1}{1 + k_a N_1} \right) \tag{17}$$

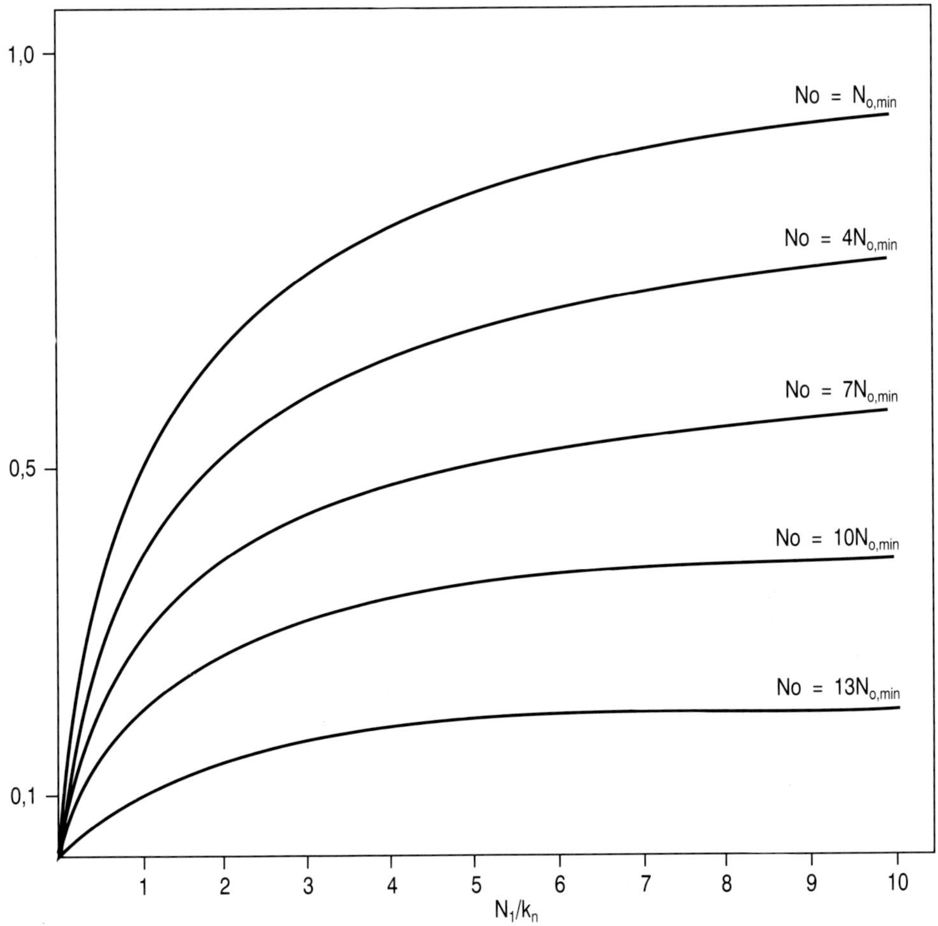

Fig. 3 Phosphorus uptake rate by algae as a function of external phosphorus concentration for various intracellular phosphorus concentrations (Eq.(16)). Assumed value of ratio $N_{o,max}/N_{o,min} = 16$.

where $N_{o,diss} = N_{o,diss}^{min} \exp(N_o/N_{o,min}-1)$ and k_a is an affinity constant between phosphate and membrane carrier. If $k_a N_{o,diss} \ll 1$, Eq.(17) is coincident with Eq.(16) (apart from the feed-back factor) once k_{N_1} is taken $=1/k_a$. Also in this connection it is worth pointing out that global eutrophication modelling generally adopts simplified solutions, which have the advantage of requiring a smaller number of constants.

Finally, the *algal mortality* rate is expressed as a bilinear function of water temperature and algal biomass concentration in the form

$$s = s_o + s_1 \cdot (T - T_{min}) \cdot B \,, \qquad (18)$$

where s_o is the mortality rate which ensures a winter steady state of algal population.

The key parameter for the mass balance of non-algal phosphorus forms is the *mineralization rate* k_{min}. Although bacterial activity is limited by the availability of dissolved oxygen and oxygen depletion results in accumulation of organic detritus at the water-sediment interface, generally only the temperature dependence of k_{min} is taken into account, by means of a relationship of the type

$$k_{min}(T) = k_{min,o} \exp (aT) \tag{19}$$

or equivalently

$$k_{min} (T) = k_{min} (T = T_o) \, \Theta^{T-T_o}, \tag{20}$$

in analogy with the temperature dependence of the algal growth coefficient.

Sources and suggestions for further reading

1. J.N. Kremer and S.W. Nixon, A Coastal Marine Ecosystem. Simulation and Analysis.
   Springer Verlag, Berlin, 1972, pp. 217
2. S.E. Jörgensen, Modelling the ecological processes, in "Mathematical Modelling of Water Quality" (G.T. Orlob, ed.).
   J. Wiley and Sons, New York, 1983, pp. 116-149.
3. J.T. Lehman, et al., The assumptions and rationales of a computer model for phytoplankton populations dynamics.
   Limnol. Oceanogr., 20, pp. 243-362, (1975)
4. V.J. Bierman, Mathematical model for selective enhancement of blue-geen algae by nutrient enrichment, in "Modelling Biochemical Processes in Aquatic Ecosystems" (R.P. Canale, ed).
   Ann Arbor Sc., 1973, pp. 1-13.

9. Parameter Evaluation and Model Calibration

Modelling lake eutrophication starts from a careful examination of the experimental framework, which characterizes the trophic stage of the lake. This suggests the choice of a *model structure* (i.e. state variables and interactions) and an acceptable *approximation* as for the description of the observed heterogeneities (e.g. the vertical box concept).

Once a reliable *numerical routine* for the solution of systems of non linear differential equations has been identified, the modeller must tackle the problem of evaluating the parameters involved in the equation coefficients.

The treatment of the problem depends on the nature of the parameter.

As has already been pointed out, there are parameters which require a proper *submodel* for their determination, such as the mixing coefficient at the depth of the thermocline m_{th} (heat transfer model, ch.5), the phosphorus release flux from sediments φ (lake-sediment interaction model, chs.4 and 7) and the altitude z_{max} of lake surface (hydrological model).

Other parameters are accessible to a direct evaluation and for these experimental investigation is considered to offer the best support to the modeller. Unfortunately, the results of laboratory experiments (e.g. algal cultures, sediment core interaction with overlying water) can hardly be extrapolated to natural ecosystems, and field investigation on a water body dominated by turbulence phenomena is conditioned by many circumstancies which tend to limit its reliability.

Nevertheless a direct experimental determination can be envisaged for the *settling velocity* of particulate matter , although with many precautions (sediment trap experiments); and experiments on starved cultures of algal species responsible for lake blooms can help in the identification of the *minimum intracellular phosphorus quota* $N_{o,min}$ for the average population in the lake.

The identification of *growth parameters* of this average algal population is usually based on “in situ” measurements of primary productivity in various temperature and light conditions, a task which is made difficult by coarse sampling and the variability of nutritional conditions. Taking into account the practical purposes of a eutrophication model once more, the evaluation of these parameters is often strictly directed towards a satisfactory reproduction of an “integral” quantity such as the average algal biomass in the lake photic zone in its seasonal evolution. *Euphotic depth* and constants determining *light extinction* are experimentally determined with photometric techniques, or even through recourse to the traditional Secchi disk (the so called “*Secchi depth*" S was found to be simply related to the total extinction coefficient k by $k = 1.8/S$).

The evaluation of the *transfer rates* between phosphorus compartments, such as mortality rate (i.e. transfer from algal to particulate non-algal), mineralization rate and uptake velocity, is much less direct. In these cases, a helpful concept is that, in a balance equation whose solution is experimentally known, one of the balance terms (that which is most difficult to evaluate directly) can be determined "a posteriori" in order to obtain the best agreement between calculated and experimental solution. The adopted value of the *control parameter* involved minimizes an expression like $\sum_i (X_{ci}-X_{mi})^2$, where X_{ci}, X_{mi} are calculated and measured value, respectively, of the state variable (the solution of the balance equation) at time t_i.

Clearly a correlation method of this type transfers onto the unknown parameter all the inaccuracies of the "directly" determined balance terms (an example of this procedure is given in Part III, ch.1).

A similar procedure can be considered as a simplified (or empirical) version of a *parameter identification* in a rigorous mathematical sense, i.e. the determination of the parameter value which minimizes a functional expression involving the solution of a differential equation, in which the unknown parameter appears. Rigorous parameter identification methods are rather complex and can probably be suggested only in the case of relatively simple systems.

The ensemble of the procedures described leads to a set of parameters which appear as more or less fitted to a particular "stage" of the lake evolution considered, as it is found from the available experimental data. For this reason they are said to constitute a model *calibration*.

The application of the model to predictive calculations concerning the future evolution of that lake, and even to "other" lakes, is only possible after careful analysis of the balance terms which have been the object of a more or less explicit fitting procedure. The extension "mutatis mutandis" of the original model to a new case is sometimes called a model *validation*.

The extent to which the variation of a parameter value affects the model results can be the subject of a systematic procedure (*sensitivity analysis*) for which general methods have been proposed. However, unless the ecosystem complexity prevents a direct check of the interdependence among the various processes, it seems convenient to concentrate attention on a few parameters of recognized "physical" meaning, and to obtain a direct (or "operational‹) experience on the model responses consequent to their variation.

A collection of parameter values adopted in lake models is presented in Appendix I.

Sources and suggestions for further reading

1. J.N. Kremer and S.W. Nixon, A Coastal Marine Ecosystem, Simulation and Analysis. Ch. 11: Sensitivity and Stability.
Springer Verlag, Berlin 1972, pp. 177-192.

2. M.B. Beck, Sensitivity analysis, calibration and validation in "Mathematical Modelling of Water Quality" (G.T. Orlob, ed.).
J. Wiley and Sons, New York, 1983, pp. 425-467

10. Modelling Eutrophic Lake Restoration

The calibration of a model on existing lake conditions (external actions including nutrient loading and sediment interaction; internal dynamics including bio-chemical processes), such as that described in the previous chapter, is an attempt at interpreting lake evolution up to the present.

The application of the calibrated model to the prediction of future lake evolution, for instance to evaluating the consequences of leaving unchanged or of modifying external (nutrient loading) conditions, is the main practical purpose of lake modelling.

There is an intrinsic limitation to the reliability of the predictive use of a model, i.e. the fact that parameters describing the lake dynamics, in particular biological and chemical processes, remain unchanged for $t > t_o$, although their evaluation is more or less affected by the type of evolution of lake state variables for $t \leq t_o$.

The time scale of a predictive calculation is generally larger than that which was necessary to interpret a seasonal evolution. For this reason many interesting statements concerning eutrophic lake restoration have been based on simple mixed reactor models which referred to the steady state total phosphorus concentration and the characteristic time of the transient term (ch. 2).

In general, predictive use of a dynamic model implies repeated calculation of annual evolutions between two consecutive total water overturns (fig.1). The difficulty of predicting the evolution of nutrient release from sediments (internal loading) suggests that the results of predictive calculations should be presented as related to various hypotheses (scenarios) as for restoration measures and sediment evolution. The conceptually simplest restoration measure is the *diversion of external loading* (N_{1ext} N_{2ext} are decreased in the model equations of Table I). The short-term effects of such an intervention are generally overevaluated by a model which does not take into account nutrient release from sediments properly. In fact, highly eutrophic lakes are characterized by an internal loading which can sustain a considerable fraction of the existing algal biomass level during many years.

An approach such as discussed in ch.4 allows us to take into account the interconnection between lake and sediment evolution, which is determined by phosphorus release from sediments, once this has been experimentally assessed. To reduce the evolution times of the system (i.e to anticipate disconnection between lake and sediments), and to lower the asymptotic phosphorus concentration, additional (internal) restoration measures must be envisaged, such as lake *water reaeration*, *artificial oxygen input*, *hypolimnetic water withdrawal*, and *sediment dredging*. In particular, **hypolimnetic**

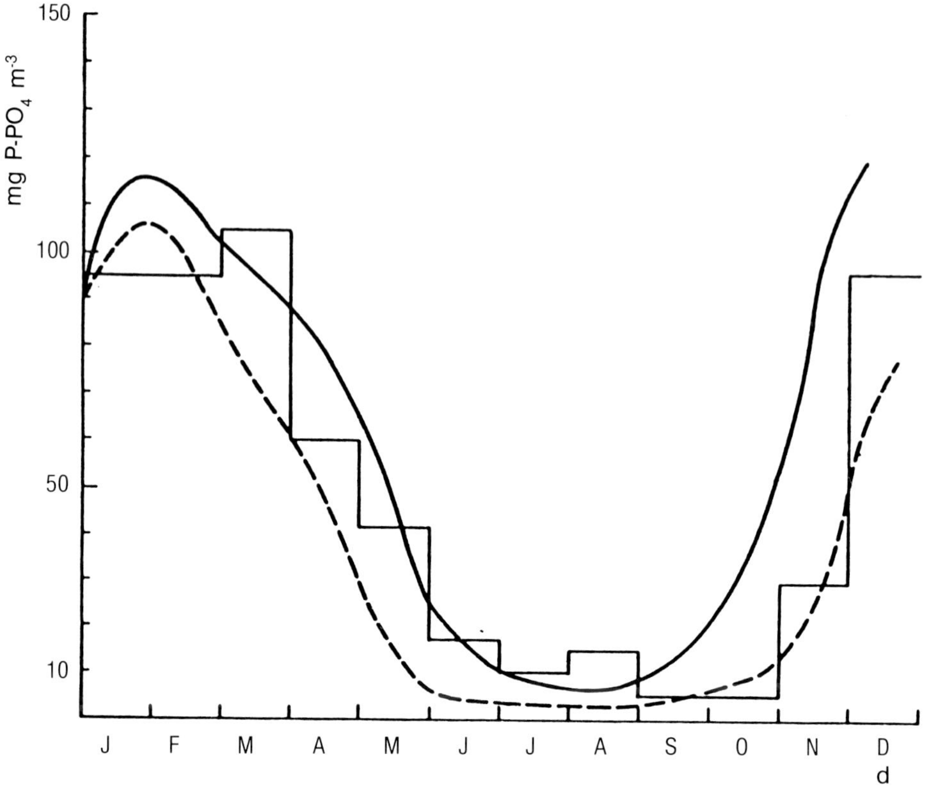

Fig. 1 Annual evolution of the average orthophosphate phosphorus concentration in the epilimnion of a bay of Lake Lugano (Agno bay). Solid line: box model results for the 1981 external P loading. Dashed line: model results for a 50% reduction of the P loading (internal loading unchanged). Histogram: experimental 1981 monthly average orthophosphate phosphorus concentrations.

water withdrawal consists of partially or totally replacing outflow at the lake surface with an equivalent outflow at hypolimnetic depths during thermal stratification. The consequence is to affect the phosphorus and oxygen balances in the hypolimnion by means of withdrawal of water, which is richer in phosphorus and poorer in oxygen than the epilimnetic waters. Obviously this measure affects the temperature distribution and vertical stability of the water column, in such a way that winter water overturn could be brought forward to autumn, when algal blooms can still be sustained. For this reason the correct design of hypolimnetic water withdrawal requires preliminary modelling analysis.

With reference to the concepts of ch.5, the initial condition is represented by the eddy diffusion coefficient profile $\chi(z,0)$ and water temperature profile $T(z,0)$ which exist before hypolimnetic water withdrawal starts. Once the new vertical distribution of outflow rate $q_{out}(t)$ has been introduced, running the heat transfer model with the "old" eddy diffusion coefficient profile supplies a new temperature profile $T(z,t_1)$. Then a new eddy diffusion coefficient profile $\chi(z,t_1)$ is obtained, for instance via Eq.(10) of ch.5

$$\chi(z,t)\left[\frac{dT}{dz}(z,t)\right]^{\frac{1}{2}} = \chi(z,0)\left[\frac{dT}{dz}(z,0)\right]^{\frac{1}{2}}, \tag{1}$$

and a recursive calculational procedure can be started.

Qualitative conclusions can be drawn from a global (advective + diffusive) hypolimnetic heat balance (fig. 2), leading to the following expression for the heating of hypolimnetic waters as a function of time:

$$(\Delta T)_{hyp} = \left(1 - e^{-\left(\frac{1}{\tau_h} + \frac{1}{\tau_{mix}}\right)t}\right)\left(T_s - T_h\right), \tag{2}$$

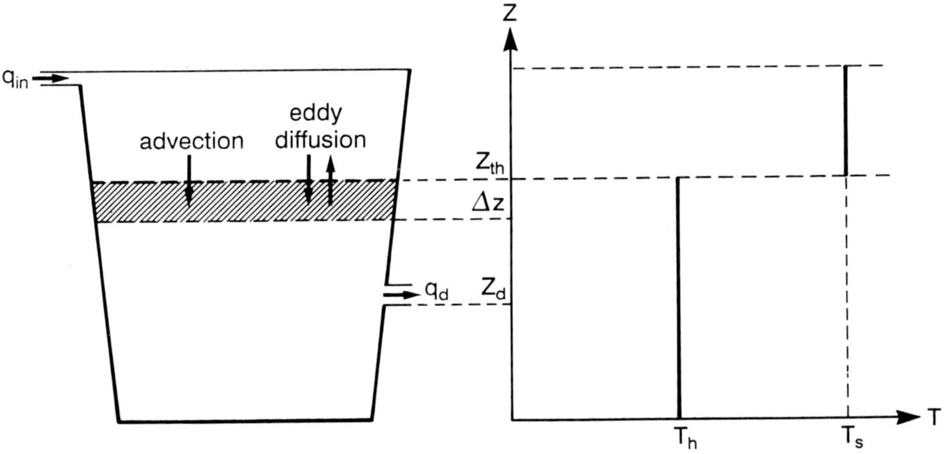

Fig. 2 Scheme of the heat transfer problem in a lake with hypolimnetic water withdrawal

where T_s, T_h are epilimnetic and hypolimnetic initial water temperatures respectively (T_s is assumed to remain unchanged);

$$1/\tau_h = q_{out}/V_{hyp}; \qquad 1/\tau_{mix} = m_{th}/h_{hyp}; \qquad h_{hyp} = V_{hyp}/A(z_{th}).$$

The increase of the mixing coefficient at the depth of the thermocline m_{th}, due to heating of hypolimnetic water as given by Eq.(2), can be estimated using Eq.(1):

$$m_{th} = m_{th}(0) \, . \, \exp\left[\left(\frac{1}{\tau_h} + \frac{1}{\tau_{mix}}\right)\frac{t}{2}\right] \tag{3}$$

The effects of hypolimnetic water withdrawal on phosphorus and oxygen mass balances are easy to evaluate provided that epilimnetic concentrations of dissolved and particulate phosphorus and dissolved oxygen N_{11}, N_{21}, Ω_1 are assumed as constant. The steady state concentrations determined by hypolimnetic outflow resulting in a hypolimnetic water mean residence time $\tau_h = V_{hyp}/q_{out}$ are as follows:

$$N_{21\infty} = N_{11} + \frac{k_{min}\,(1/\tau_{sed})}{(1/\tau_h + 1/\tau_{mix})\,(k_{min} + 1/\tau_h + 1/\tau_{sed})}\,N_{12} + \frac{\varphi_2}{h_{hyp}\,(1/\tau_h + 1/\tau_{mix})}$$

$$N_{22\infty} = \frac{1/\tau_h + 1/\tau_{sed}}{k_{min} + 1/\tau_h + 1/\tau_{sed}}\,N_{21} \tag{4}$$

$$\Omega_{2\infty} = \Omega_1 - \frac{k_\Omega\,k_{min}\,N_{22\infty} + (\varphi_{\Omega 2}/h_{hyp})}{1/\tau_h + 1/\tau_{mix}}$$

where $\tau_{sed} = h_{hyp}/v_s$ and the other symbols were defined in ch.7.

The reduction of the effect of internal loading on hypolimnetic phosphorus concentration, as well as the reduction of the volume source of dissolved phosphorus due to mineralization, are exhibited by the term $1/\tau_h$ at the denominators in Eqs.(4).

Balance equations in the framework of a box approximation allow us to perform predictive calculations also on the effect of lake water reaeration and artificial oxygen input into the hypolimnion.

Nevertheless, a lot of empirical informations enter into the modelling of a reaeration or oxygenation experiment, as for instance the hypolimnetic oxygen consumption rate and the dependence of phosphorus release on the dissolved oxygen concentration at the water-sediment interface.

Once a realistic hypothesis has been formulated for the restoration measures to envisage, the evaluation of the lake recovery time relys principally upon the assessment of the time required to achieve disconnection between lake and sediment.

Apart from modelling effort, a statistical investigation on many eutrophic lakes, which had undergone a drastic reduction of the external phosphorus loading without internal measures, was able to assess a *delay time* before internal loading begins to decrease (~ 5 years).

After disconnection from the sediment, the lake evolution is essentially determined by the lake parameters, in particular by the water renewal time, and recovery can be easily predicted, provided that the steady state nutrient concentration is compatible with the adopted water quality criteria.

An example of predictive calculation of the recovery time of a eutrophic lake is given in Part III, ch.2.

Sources and suggestions for further reading

1. R.Gächter and D.M. Imboden, Lake restoration, in "Chemical Processes in Lakes" (W. Stumm, ed.).
   J. Wiley and Sons, New York, 1984, pp. 365-388.
2. Lake Restoration by Reduction of Nutrient Loading (H. Sas, coord.).
   Academia Verlag, 1989, pp. 497.

Appendix

Parameter Values for Lake Eutrophication Models

1. Algal growth

$$\mu_{max} = a \exp(bT) \quad (5\,^\circ C \leq T \leq 25\,^\circ C)$$
$$a = 0.59\,d^{-1};\ b = 0.0633\,^\circ C^{-1} \tag{1}$$

$$\mu_{max} = \mu_{max}(T = 20\,^\circ C)\ \Theta^{T-20}$$
$$\mu_{max}(T = 20\,^\circ C) = 1\text{–}2\,d^{-1} \tag{2}$$

$$(T = 25\,^\circ C) = 0.8\text{–}2.1\,d^{-1} \tag{3}$$

$$1.02 < \Theta < 1.06 \tag{2}$$

$$1.07 < \Theta < 1.10 \tag{3}$$

$$\theta(z,t) = I(z,t)/I_{opt} \exp\left[1 - I(z,t)/I_{opt}\right]$$
$$I_{opt} = 0.27\,\text{kcal m}^{-2}\text{min}^{-1} \tag{4}$$

$$= 0.3\,\text{kcal m}^{-2}\text{min}^{-1}\ \text{(Blue–greens)} \tag{5}$$

$$f(I) = I/(k_I + I)$$
$$k_I = 0.18\,\text{kcal m}^{-2}\text{min}^{-1} \tag{2}$$

$$g(No) = 1 - N_{o,min}/N_o$$
$$N_{o,min} = 0.107 - 0.215\ \text{(mol P/1 cell volume)} \tag{3}$$

$$= 0.001 \tag{6}$$

$$u = u_{max} \cdot \frac{N_{o,max} - N_o}{N_{o,max} - N_{o,min}} \cdot N_1/(k_N + N_1)$$

$u_{max} = 0.024 - 0.133\ d^{-1}$ (3)

$= 0.0014 - 0.003\ d^{-1}$ (6)

$N_{o,max} = 0.8 - 16 \times 10^{-8}$ (μmoles cell^{-1}) (5)

$= 0.013$ (6)

$k_N = 0.02$ g m -3 (2,5)

2. Algal mortality

$s = s_o + s_1 (T - T_{min}) B$

$s_1 = 0.0015\ (°C\ g\ m^{-3}\ d)^{-1}$ (3)

3. Organic phosphorus mineralization

$k_{min} = k_{min} (T = 10\ °C)\ \Theta^{T-10}$ (6)

$\Theta = 1.072$ (6)

$k_{min} = 0.07 - 0.1\ d^{-1}$ (epilimnion)

$= 0.02$ (hypolimnion) (7)

4. Settling velocity of particulate matter

$v_s = 0.2\ m\ d^{-1}$ (2)

$= 0.15 - 0.4\ m\ d^{-1}$ (3)

5. Light extinction coefficient

$k = k_o + k_1 B$

$k_o = 0.27\ m^{-1}$ (2)

$= 0.633\ m^{-1}$ (3)

$k_1 = 0.18\ (mg\ d.w.\ m^{-3})^{-1}\ m^{-1}$ (2)

$= 0.17\ (mg\ d.w.\ m^{-3})^{-1} m^{-1}$ (3)

6. Eddy diffusion coefficient

At the thermocline, during stagnation:

$\chi\ (z_{th}) = 0.02 - 0.04\ cm^2 s^{-1}$ (7)

in the hypolimnion, during stagnation:

$\chi = 0.05 - 0.15\ cm^2 s^{-1}$ (7)

References of the Appendix

1. R.W. Eppley, "Temperature and phytoplankton growth in the sea", Fish. Bull., 70, 1063-1085 (1972).
2. C.W. Chen and G.T. Orlob, "Ecological simulation for aquatic environments", System Analysis and Simulation in Ecology, Vol.3 (B.C. Patten, ed.), Academic Press, 1975, 476-588.
3. V.J. Bierman, "Mathematical model of the selective enhancement of blue-green algae by nutrient enrichment", Modelling Biochemical Processes in Aquatic Ecosystems, (R.P. Canale, ed.) Ann Arbor Sc., 1976, pp.1-31.
4. R.A. Parker, "Some problems associated with computer simulation of an ecological system", Mathematical Theory of the Dynamics of the Biological Populations, Oxford Univ., 1973, 269-288.
5. J.T. Lehman, D.T. Botkins and G.E. Likens, "The assumptions and rationales of a computer model for phytoplankton population dynamics", Limnol. Oceanogr., 7, 343-362 (1975).
6. S.E. Jörgensen, "A eutrophication model for a lake", Ecol. Modell., 2, 147-165 (1976).
7. D.M. Imboden and R. Gächter, "A dynamic model for trophic stateprediction", Ecol. Modell., 4, 77-98 (1978).

A wide collection of values of ecological parameters is given by M. Straskraba and A.H. Gnauck, Freshwater Ecosystems, Modelling and Simulation, Elsevier, 1985, pp.309.

Acknowledgment

Material presented in Part I is partly reprinted from the author's chapter "Modeling Lake Eutrophication" in Encyclopedia of Fluid Mechanics, Volume 10, Surface and Groundwater Flow Phenomena, Nicholas P. Cheremisinoff, Editor. Copyright c 1990 by Gulf Publishing Company, Houston, Texas. Used with permission.

Part II

MODELLING HEAVY METAL POLLUTION OF LAKES

I. Phenomenology of Chemical Pollution of Lakes and General Modelling Concepts

Human activities imply to discharge into the lake chemical substances other than algal nutrients, such as pesticides used in agriculture and heavy metals from factories.

In the same way as nutrients, these substances are present in lake water under various chemical forms (ions, radicals), and each of these in liquid (dissolved) and in solid phase (being adsorbed on suspended particulate matter).

Chemical substances which are adsorbed on individuals of lake biota can have toxic effects on it, and interfere with the eutrophication process. On the other hand adsorption associates pollutants to settling materials, and eventually to lake sediments, from which they are remobilized only in very particular conditions.

Dissolved substances, which are ready to be adsorbed and potentially toxic, participate in water movements, i.e. they undergo advective and diffusive (turbulent) transport, in particular they are flushed by outflowing waters.

In a general way a pollutant distribution model can be considered as a mathematical tool to link the pollutant concentration in lake to that in inflowing waters. Once the toxic threshold of the pollutant with respect to a particular biologic population is established, the model enables us to evaluate the highest pollutant concentration in inflowing waters which is compatible with the life of this population in the lake waters.

The balance equation for a substance, which is not necessarily connected with algal growth, such as a generic chemical pollutant, has obviously the same (advective and diffusive) transport terms as the balance equation for a nutrient, if this equation is related to the same lake region.The terms which are specific of the mass balance of a chemical pollutant, are those related to chemical reactions with other dissolved substances, as well as other processes affecting its concentration, such as volatilization, photolysis, hydrolysis, etc.

A particular attention must be payed to adsorption-desorption processes of the pollutant by matter suspended in the water body. In a model of the trophic state of a lake (phosphorus mass balance), this process was taken into account only in the sediments, where it was described by means of the concept of "linear equilibrium adsorption isotherm" between orthophosphate dissolved in pore water and phosphorus adsorbed on solid sediment particles (see Part I, ch.4).

In the water body, the existence of a compartment "inorganic particulate phosphorus" and the exchange with the dissolved phase were neglected, mineralization of organic particulate phosphorus and uptake of inorganic dissolved phosphorus by phytoplankton being the only interactions considered.

The linear isotherm concept can be generalized to the phases of any chemical substance, as well as that of distribution (or partition) coefficient. Obviously, a simple proportionality relation between pollutant concentration in solid particles (mg g^{-1}) and concentration of its dissolved phase in water (mg m^{-3}) allows important formal simplifications and is extensively adopted in ecological models. The proportionality constant (distribution coefficient, m^3 g^{-1}) is dependent on the physico-chemical properties (temperature, pH, redox potential) of water. Its determination is a key point in every dispersion model of a chemical pollutant in a water body, where suspended matter is present.

Every chemical pollutant requires a specific modelling analysis, i.e. the adoption of proper simplifications taking into account its reactivity and the characteristic times of the processes, which affect its concentration.

However, general discussions can be found in the literature, which are devoted to a class of pollutants, such as hydrophobic organic pollutants or heavy metals.

In the following chapters the case of heavy metals is treated, which is the simplest one, especially if one refers to the total (all ionic forms) metal concentration as to the state variable in a given lake region.

2. Heavy Metal Distribution in the Lake-Sediment System

Release of soluble forms of a chemical pollutant from lake sediments has not revealed, generally speaking, as important from an ecological point of view, as phosphorus release in eutrophic lakes. However the modelling analysis of this phenomenon is useful in order to put in evidence the quantities which determine its possible influence on the lake evolution.

Aiming at predicting the time evolution of the average total metal (all species) total concentration (dissolved+solid phase), a lacustrine ecosystem can be considered as composed either by two permanently separated, interacting systems, i.e. lake and sediment, or by two vertically superposed lake regions, i.e. epilimnion and hypolimnion, the sediment action on the lake being described by a source term,whose intensity is given by separate considerations.

As for the system lake-sediment, the state of the lake and that of the sediment are each described by means of a single variable, i.e. the lake is considered as a homogeneously mixed reactor: thermal stratification is taken into account only by means of the effective water renewal time concept. The distribution of suspended matter is supplied either by experimental investigations, or by a separate trophic state model.

The mathematical structure of the model is given by a system of two linear differential equations with constant coefficients for two unknown variables: the (total) metal concentrations in lake C_L and sediment C_S.
The solutions of the system of equations have the following form:

$$y_i = A_{1i} \exp(-\alpha_1 t) + A_{2i} \exp(-\alpha_2 t) + y_{i\infty} \quad (i = 1,2) \tag{1}$$

The time constants α_1, α_2 are expressed in terms of the parameters characterizing the intrinsic properties of the system; the coefficients A_{ij} are obtained having recourse to the initial conditions $y_i(0) = y_{io}$, and the asymptotic terms $y_{i\infty}$ involve the source intensities.
If C_p gives the concentration of particulate matter (g d.w. m^{-3}), and C_{ads} the concentration of the metal inside the particulate matter (mg g^{-1}), the total concentration of the metal (mg m^{-3}) is given by:

$$C_t = C_d + C_{ads} \cdot C_p$$

where C_d is the concentration of dissolved metal.
The key parameters of the model are the distribution constants, which determine the equilibrium conditions between the adsorbed metal phase and the metal dissolved in surrounding water:

$$C_{L,ads} = k_p C_{Ld} \qquad C_{S,ads} = k_{ps} C_{Sd} \, ,$$

where C_{Sd} is the dissolved metal concentration in pore-water. The constants k_p , k_{ps} ($m^3\ g^{-1}$) depend on the temperature and pH of the water region considered.
As direct consequence of the above presented statements, the following relationships are obtained:

$$C_{Ld} = f_p C_{Lt} \quad C_{Lp} = C_{S,ads} \cdot C_p = (1 - f_p)\, C_{Lt} \, ,$$

where $f_p = 1/(1 + k_p C_p)$.

With the assumptions that exchange of metal between lake and sediments only takes place via sedimentation of particulate matter and net release (i.e. release minus reprecipitation) of dissolved metal from pore-water; that diffusion from the *active sediment layer* (i.e. the layer affected by water turbulence and bioturbation) to permanent sediment is negligible,and that the solid sediment mass in the active layer is constant (fig.l), the balance equations for the total metal in the system lake+sediment can be written as follows:

$$d/dt\, C_{Lt} = -\left[k_{out} + k_{sed}\,(1 - f_p)\right] C_{Lt} + v_r p/(h k_{ps})\, C_{Sads} + k_{in} C_{ext}$$

$$h_s \left[p/k_{ps} + (1 - p)\, \rho_s\right] d/dt\, C_{Sads} = h k_{sed}\,(1 - f_p)\, C_{Lt} - (v_r p/k_{ps} + r_s)\, C_{Sads} \qquad (2)$$

where:

$k_{out} = q_{out}/V_L$ — q_{out} = outflow rate ($m^3\ yr^{-1}$)
V_L = lake volume (m^3)

$k_{sed} = v_s/h$ — v_s = settling velocity of particulate matter ($m\ yr^{-1}$)
h = mean lake depth (m) = V_L/A

A = lake surface = active sediment surface area (m^2)
$k_{in} = q_{in}/V_L$ — q_{in} = inflow rate ($m^3\ yr^{-1}$)
C_{ext} = total metal concentration in inflowing water ($mg\ m^{-3}$)
v_r = release "velocity" of dissolved metal from pore-water($m\ yr^{-1}$)
release flux $\varphi_r = v_r A p C_{Sads}/k_{ps}$
p = average porosity in the active sediment layer

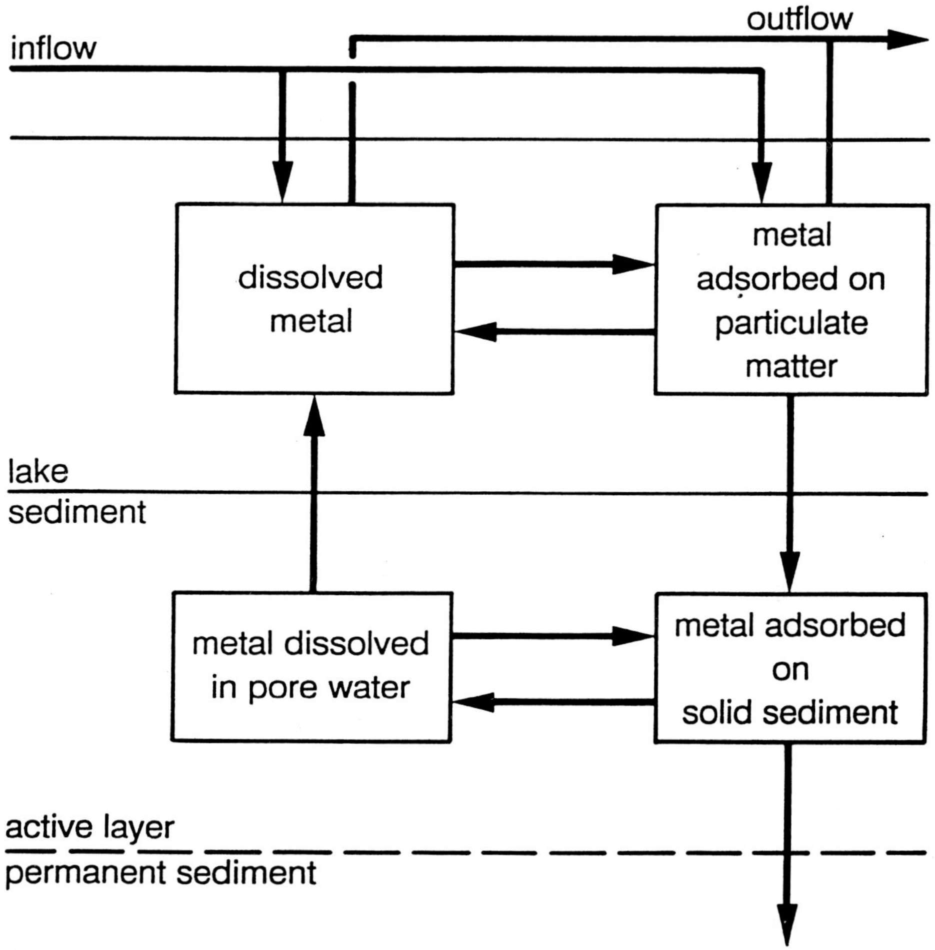

Fig. 1 Outline of the modelled lake-sediment system

ρ_s = density of solid sediment (g m $^{-3}$)
h_s = thickness of the active sediment layer (m)
The solid sediment mass in the active sediment layer is given by

$$M = \rho_s A (1-p) h_s$$

and the time to renew this layer is given by

$$\tau_s = h_s \rho_s (1-p)/r_s.$$

Note that it is more convenient to refer the metal mass balance in the sediment to the metal concentration inside the solid sediment (C_{Sads}) instead of the total concentration C_{St}.

The coefficients of the balance equations are assumed to be constant during the period to which the calculations refer. Putting:

$a_{11} = k_{out} + k_{sed}(1-f_p)$ $a_{12} = v_r p/(hk_{ps})$ $a_{10} = k_{in}C_{ext}$

$a_{21} = hk_{sed}(1-f_p)/N_s$ $a_{22} = (v_r p/k_{ps} + r_s)/N_s$

where $N_s = h_s[p/k_{ps} + (1-p)\rho_s]$,

time constants and asymptotic concentrations are given by:

$$\alpha_1(\alpha_2) = 1/2\left[a_{11} + a_{22} + (-)\sqrt{(a_{11}-a_{22})^2 + 4a_{12}a_{21}}\right]$$

$$C_{Lt\infty} = a_{22}a_{10}/(a_{11}a_{22} - a_{12}a_{21}) \qquad C_{Sads\infty} = a_{21}C_{Lt\infty}/a_{22}.$$

The explicit expression for the time constants is as follows:

$$\alpha_1(\alpha_2) = \frac{1}{2}\left\{k_{out} + k_{sed}(1-f_p) + \frac{\frac{v_r p}{k_{ps}} + r_s}{h_s\left[\frac{p}{k_{ps}} + (1-p)\rho_s\right]} + \right.$$

$$\left. + (-)\sqrt{\left[k_{out} + k_{sed}(1-f_p) - \frac{\frac{v_r p}{k_{ps}} r_s}{h_s\left[\frac{p}{k_{ps}} + (1-p)\rho_s\right]}\right]^2 + 4\frac{v_r p}{k_{ps}}\frac{k_{sed}(1-f_p)}{h_s\left[\frac{p}{k_{ps}} + (1-p)\rho_s\right]}}\right\} \qquad (3)$$

The explicit expressions for the asymptotic concentrations are as follows:

$$C_{Lt\infty} = \frac{k_{in}C_{ext}}{k_{out} + \frac{r_s}{\frac{v_r p}{k_{ps}} + r_s}k_{sed}(1-f_p)} \qquad (4)$$

$$C_{Sads\infty} = \frac{hk_{sed}(1-f_p)}{k_{out}\left(\frac{v_r p}{k_{ps}} + r_s\right) + k_{sed}(1-f_p)r_s}k_{in}C_{ext} \qquad (5)$$

The evolution curves $C_{Lt}(t)$, $C_{Sads}(t)$ involve both time constants α_1, α_2, as well as the total decrease of both lake and sediment metal concentrations $C_{Lt}(0)$-$C_{Lt\,\infty} = \Delta C_L$,

$C_{Sads}(0)$-$C_{sads\infty} = \Delta C_{Sads}$: the explicit expressions for the coefficients A_{ij} of eq. (l) are given in App.I.

If the release velocity v_r vanishes (vanishing feedback from sediments to metal concentration in lake), these curves have the following simple form:

$$C_{Lt}(t) = C_{Lt}(0) \exp(-\alpha_1 t) + \left[1 - \exp(-\alpha_1 t)\right] C_{Lt\infty} \tag{6}$$

$$C_{Sads}(t) = C_{Sads}(0) \exp(-\alpha_2 t) + \left[1 - \exp(-\alpha_2 t)\right] C_{Sads\infty} \tag{7}$$

i.e. the time constants α_1, α_2 determine separately the evolution of metal concentration in lake and sediment respectively.

The limit values of quantities (3), (4), (5) when v_r tends to zero are as follows:

$$\alpha_1 (v_r = 0) = k_{out} + k_{sed}(1 - f_p) \tag{3'}$$

$$\alpha_2 (v_r = 0) = \frac{r_s}{h_s\left[\dfrac{r}{k_{ps}} + (1-p)\,\rho_s\right]} \tag{3''}$$

$$C_{Lt\infty}(v_r = 0) = \frac{k_{in}\, C_{ext}}{k_{out} + k_{sed}(1 - f_p)} \tag{4'}$$

$$C_{Sads\infty}(v_r = 0) = \frac{h k_{sed}(1 - f_p)}{r_s\left[k_{out} + k_{sed}(1 - f_p)\right]} k_{in}\, C_{ext} \tag{5'}$$

The influence of lake-sediment interaction on the asymptotic total metal in-lake concentration is expressed by the factor $r_s/(r_s + v_r p/k_{ps})$ in the denominator at the r.h.s. of eq. (4). This factor represents the ratio between the metal flux downwards and the total (upwards + downwards) flux from the active sediment layer.

The influence of sediment interaction on the time constants of the lake is expressed by the second term under square root in eq.(3). This term is proportional to the product of the total metal release flux referred to the unit lake volume times the total sedimentation flux referred to the unit metal mass in the solid sediment of the active layer.

3. Heavy Metal Distribution in the Epilimnion-Hypolimnion System

Recourse to the modelling framework presented in ch.2 is made mandatory by the experimental observation that net release of metal from sediment affects metal concentration in lake to such an extent that the interconnection between lake and sediment evolution cannot be disregarded.

If this is not the case, i.e. if v_r/k_{ps} is negligible with respect to r_s, the same mathematical apparatus can be applied to the treatment of separate total metal mass balances in epilimnion and hypolimnion, a possible metal release from sediment being discussed via a time independent source term in the hypolimnion balance equation.

To validate such a model experimental information is needed on metal distribution constant and particulate matter concentration in epilimnion and hypolimnion separately, as well as on the mixing coefficient across the thermocline.

The state variables are now the total metal concentrations in epilimnion and hypolimnion, C_{et} and C_{ht} respectively. Dissolved metal exchanges between the two regions are determined by the mixing coefficient at the depth of the thermocline m_{th} (m yr $^{-1}$) (fig. 2).

In reality, both thermocline depth (i.e. epilimnion volume) and thermocline stability (i.e. mixing coefficient) undergo a seasonal evolution. In a metal distribution model with time scale in years (whose applications are limited to long term predictions), this is only roughly represented by adopting a fixed thermocline depth at its average summer value, and a mixing coefficient at its yearly average value.

Assuming that the net release flux from sediments is negligible, the balance equations for the total metal in the system epilimnion+hypolimnion are as follows:

$$d/dt\, C_{et} = -\left[k_{out,e} + k_{mix,e} f_{pe} + k_{sed,e}(1 - f_{pe})\right] C_{et} + k_{mix,e} f_{ph} C_{ht} + k_{in,e} C_{ext}$$

$$d/dt\, C_{ht} = \left[k_{mix,h} f_{pe} + k_{sed,h}(1 - f_{pe})\right] C_{et} - \left[k_{mix,k} f_{ph} + k_{sed,h}(1 - f_{ph})\right] C_{ht} \qquad (1)$$

where:

$k_{out,e}$	=	q_{out}/V_e	V_e = epilimnion volume	
$k_{mix,e}$	=	m_{th}/h_e	$h_e = V_e/S_{th}$	S_{th} = thermocline area
$k_{sed,e}$	=	v_s/h_e	$f_{pe} = 1/(1 + k_{pe} C_{pe})$	
k_{pe}	=	distribution constant in the epilimnion		

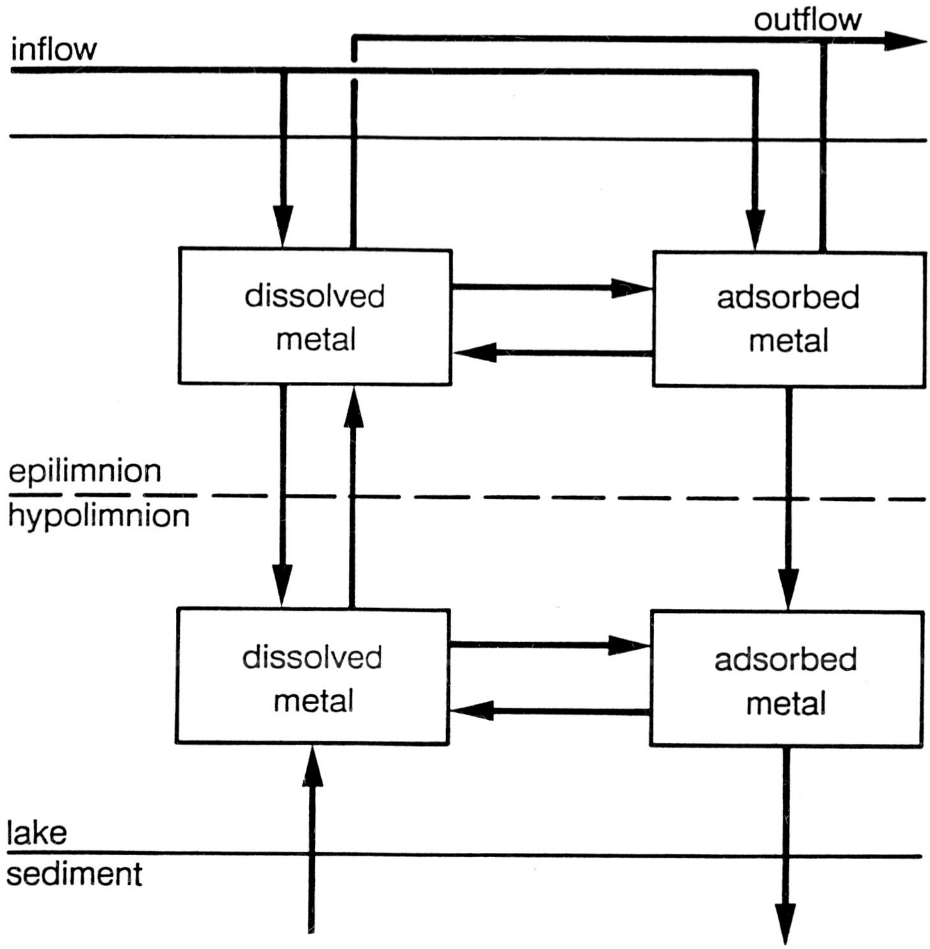

Fig. 2 Outline of the modelled epilimnion-hypolimnion system

C_{pe}	=	concentration of particulate matter in the epilimnion		
$k_{in,e}$	=	q_{in}/V_e		
$k_{mix,h}$	=	m_{th}/h_h	V_h = hypolimnion volume	$h_h = V_h/S_{th}$
$k_{sed,h}$	=	v_s/h_h		
k_{ph}	=	distribution constant in the hypolimnion		
C_{ph}	=	concentration of particulate matter in hypolimnion.		
f_{ph}	=	$1/(1+ k_{ph}\, C_{ph})$		

In order to utilize the general expressions for time constants and asymptotic values of the state variables (ch.2), the following coefficients are defined:

$$a_{11} = k_{out,e} + k_{mix,e}\, f_{pe} + k_{sed,e}\,(1 - f_{pe}) \qquad a_{12} = k_{mix,e}\, f_{ph}$$
$$a_{10} = k_{in,e}\, C_{ext}$$

$$a_{21} = k_{mix,h}\, f_{pe} + k_{sed,h}\,(1 - f_{pe}) \qquad a_{22} = k_{mix,h}\, f_{ph} + k_{sed,h}\,(1 - f_{ph})\,.$$

The time constants of the lake evolution are given by:

$$\alpha_1\,(\alpha_2) = 1/2 \Big\{k_{out,e} + k_{mix,e}\, f_{pe} + k_{sed,e}\,(1 - f_{pe}) + k_{mix,h}\, f_{ph} + k_{sed,h}\,(1 - f_{ph})$$
$$+ k_{sed,h}\,(1 - f_{ph}) \pm \sqrt{\Big[k_{out,e} + k_{mix,e}\, f_{pe} + k_{sed,e}\,(1 - f_{pe}) - k_{mix,h}\, f_{ph} -}$$
$$\overline{- \big\{k_{sed,h}\,(1 - f_{ph}\,)\big]^2 + 4k_{mix,h}\, f_{ph}\Big[k_{mix,h}\, f_{pe} + k_{sed,h}\,(1 - f_{pe}\,)\Big]}\Big\}\,. \tag{2}$$

The asymptotic total metal concentration in the epilimnion is given by:

$$C_{et\infty} = \frac{k_{in,e}\, C_{ext}}{k_{out,e} + k_{mix,e}\, f_{pe} + k_{sed,e}\,(1 - f_{pe}) - k_{mix,e}\, f_{ph}\, \dfrac{k_{mix,h}\, f_{pe} + k_{sed,h}\,(1 - f_{pe}\,)}{k_{mix,h}\, f_{ph} + k_{sed,h}\,(1 - f_{ph})}} \tag{3}$$

The asymptotic total metal concentration in the hypolimnion is given by:

$$C_{ht\infty} = C_{et\infty}\, \frac{k_{mix,h}\, f_{pe} + k_{sed,h}\,(1 - f_{pe})}{k_{mix,h}\, f_{ph} + k_{sed,h}\,(1 - f_{ph})} \tag{4}$$

The influence of the epilimnion-hypolimnion interaction on the asymptotic metal distribution along the water column is obviously determined by $k_{mix,e}$, while that on the time constants of the system evolution is determined by the magnitude of the second term under square root in eq. (2).

Even in lakes, which are stably stratified during summer, this term is never negligible on a annual basis, in such a way that the long term evolution of the metal in-lake concentration is principally determined by the property of the lake as a mixed reactor.

The effect of a possible non vanishing net release flux from bottom sediments can be evaluated by introducing a term

$$a_{20} = \varphi_r / h_h ,$$

where φ_r represents the dissolved metal flux, into the r.h.s. of the second eq. (1). Then, the ratio $C_{et\infty}/C_{ht\infty}$ is given by

$$\frac{C_{et\infty}}{C_{ht\infty}} = \frac{a_{22}\, a_{10} + a_{12}\, a_{20}}{a_{21}\, a_{10} + a_{11}\, a_{20}} =$$

$$= \frac{\left[k_{mix,h}\, f_{ph} + k_{sed,h}\, (1 - f_{ph})\right] k_{in,e}\, C_{ext} + k_{mix,e}\, f_{ph}\, \dfrac{\varphi_r}{h_h}}{\left[k_{mix,h}\, f_{pe} + k_{sed,h}\, (1 - f_{pe})\right] k_{in,e}\, C_{ext} + \left[k_{out,e} + k_{mix,e}\, f_{pe} + k_{sed,e}\, (1 - f_{pe})\right] \dfrac{\varphi_r}{h_h}} \quad (5)$$

The evolution curves $C_{et}(t)$, $C_{ht}(t)$ are expressed in terms of the concentration variations $\Delta C_{et} = \Delta C_{et}(0) - C_{et\infty}$; $\Delta C_{ht} = C_{ht}(0) - C_{ht\infty}$, with the coefficients A_{ij} of eq. (1) in ch.2 as explicitly given in App.II.

It is worth pointing out that *short term predictive calculations*, i.e. calculations concerning the variation of state variables at times of the same order of magnitude as the characteristic times of the involved processes, need the solution of balance equations with time dependent coefficients, hence numerical integration procedures. The time scale of these calculations is in months at the most, and the seasonal evolution of quantities such as the concentration of suspended matter C_p and the mixing coefficient m_{th}, as well as variations in the physico-chemical properties of the water body or in the metal load, can be taken into account. A short description of a computer code, written for this purpose, is given beneath. The balance equations of the system solved by the code are referred to the epilimnion (region 1) and to the hypolimnion (region 2) of a stratified lake respectively, whose separation surface (thermocline) is placed at a fixed depth (the average depth during summer stratification) and is characterized by a mixing coefficient underlying a seasonal evolution.

The system, which is integrated by an appropriate numerical solution routine, is as follows:

```
DY(1)  =  -F11*Y(1)  +  F12*Y(2)  +  F10
DY(2)  =   F21*Y(1)  -  F22*Y(2)  +  F20
```

with:

```
F11  =  KOUT + KMIX1*FP1 + KSED1*(1.-FP1)
F12  =  KMIX1*FP2
```

```
F10  =  KINR*CTR
F21  =  KMIX2*FP1 + KSED2*(1.-FP1)
F22  =  KMIX2*FP2 + KSED2*(1.-FP2)
F20  =  PSI

KOUT   =  QOUT/VOL1
KMIX1  =  STH*MTH/VOL1
KSED1  =  STH*VS1/VOL1
FP1    =  1./(1.+KP1*CP1)
FP2    =  1./(1.+KP2*CP2)
KINR   =  QIN/VOL1
KMIX2  =  STH*MTH/VOL2
KSED2  =  STH*VS2/VOL2
PSI    =  STH*PHI/VOL2
```

The parameters, which appear in the last group of statements, are partly given by a single value (VOL1; VOL2; STH = area of the lake section at the depth of the thermocline; VS1 = settling velocity of suspended matter in epilimnion; KP2 = distribution constant in the hypolimnion; VS2 = settling velocity of suspended matter in hypolimnion; PHI = yearly average release flux from sediments);and partly given each by a vector of 12 average monthly values (QIN = QOUT = inflow-outflow rate; CTR = total metal concentration in the inflowing water; MTH = mixing coefficient; KPl = distribution constant in the epilimnion; CP1 = suspended matter concentration in the epilimnion; CP2 = suspended matter concentration in the hypolimnion).

Input data include integration data for the numerical routine, the number of month NMONTH of the considered evolution, and the initial values Y(1), Y(2) of the total metal concentration in epilimnion and hypolimnion respectively.

4. The Relationship Between External Metal Load and Metal Concentration in Solid Sediment

An interesting problem in the field of heavy metal pollution of lakes is to evaluate former pollutant input rates from pollutant concentration profiles in sediment cores. This problem was solved in the following assumptions:

1) the metal concentration in dry solid just sedimented matter is the same as in sedimenting ("suspended" in lake) matter. Using concepts and symbols of ch.2, the latter concentration is related to the total metal concentration in lake by the relationship

$$C_{L,ads} = \left[k_p/(1 + k_p\, C_p)\right] C_{Lt}\, ; \tag{1}$$

2) the distance z from the sediment surface at time t of matter, which was just sedimented ($z = 0$) at time t_o depends only on the accumulation rate s ($m\ yr^{-1}$) (metal transport processes inside the sediment are neglected). If s is constant the relationship is simply

$$z = s\,(t - t_o)\,. \tag{2}$$

Taking into account that to C_{Lt} increasing with time corresponds $C_{S,ads}$ decreasing with increasing z, the following relationships are obtained:

$$C_{S,ads}\,(z) = C_{L,ads}\,(t_o) = \left[k_p/(1 + k_p\, C_p)\right] C_{Lt}\,(t_o) \tag{3}$$

$$- d/dz\, C_{S,ads} = (dt/dz)\, d/dt\, C_{S,ads} =$$

$$= (1/s)\, d/dt \left[k_p\, C_{Lt}/(1 + k_p\, C_p)\right]. \tag{4}$$

Having recourse to the balance equation for the total metal in the lake, written in the form

$$d/dt\, C_{Lt} = -(1/\tau)C_{Lt} + I(t),$$

where

$$1/\tau = 1/\tau_w + 1/\tau_{sed}\,,$$

τ_w is the mean residence time of water and $\tau_{sed} = [k_{sed}\,(1 - f_p)]^{-1}$ is the characteristic time of the metal sedimentation process; I(t) is the metal load per unit lake volume (mg m^{-3} yr^{-1}), it is easy to obtain the relationship betwen metal load I(t) and metal concentration $C_{S,ads}(z)$ [briefly C(z)] in the solid sediment:

$$I(t) = (1/k_p + C_p) \Big[- s\, d/dz\, C(z) + (1/\tau)\, C(z) - $$
$$- (1/k_p + C_p)\, C(z)\, d/dt \left[k_p/(1 + k_p C_p) \right] \Big] . \qquad (5)$$

Assuming k_p and C_p as time constant, eq. (5) becomes

$$I(t) = (1/k_p + C_p) \left[-s\, d/dz\, C + (1/\tau)\, C(z) \right] . \qquad (6)$$

Furthermore, if measurements of C (z = 0), repeated for many years, show that the metal concentration at the surface of the sediment does not vary, the assumption C = const (z) in the first few cm of the sediment core is justified, and eq. (1) gives simply:

$$I = (1/k_p + C_p)\, (1/\tau)\, C . \qquad (7)$$

An attempt at evaluating the average metal load when the metal content of sediment is varying with depth requires the integration of eq. (1) between the "present" time $t = t_o + z_o/s$ and time t_o, when sediment now at depth z_o was at the sediment surface. The following equation is obtained:

$$[I] = \left[1/(z_o/s) \right] (1/k_p + C_p) \left[C(0) - C(z_o) \right] + (1/\tau)\, (1/k_p + C_p)\, [C] , \qquad (8)$$

where:

[I] = average metal load mg m^{-3} yr^{-1} during the considered period of z_o/s years
[C] = average metal concentration in dry sediment mg g^{-1} in the (0, z_o) layer of the sediment core.

In Part III concepts and methods described in these chapters are applied to the problem of the evolution of Cu concentration in a subalpine lake which has suffered heavy industrial pollution since the Twenties (Lake Orta).

Sources and suggestions for further reading

1. D.M. Imboden and R.P. Schwarzenbach, Spatial and temporal distribution of chemical substances in lakes: modelling concepts, in "Chemical Processes in Lakes" (W.Stumm, ed.).
   J. Wiley and Sons, New York, 1984, pp. 1-30

2. D.M. Imboden, et al., Die Rekonstruktion früherer Stofffrachten in einem See mittels Sedimentuntersuchungen.
   Schweiz. Zeits. Hydrol. 42/1, 1-14 (1980)

3. S.E. Jörgensen, Modeling the Distribution and Effect of Toxic Substances in Rivers and Lakes, in "Mathematical Modeling of Water Quality: Streams, Lakes and Reservoirs" (G.T. Orlob, ed.).
   J. Wiley and Sons, New York, 1983, pp. 395-424

Appendix I

Coefficients Aij of Eq.(1) in Ch.2 for the System Lake-Sediment

$$A_{11} = \frac{\frac{h}{N_s} k_{sed} (1 - f_p)\, \Delta C_L - \left[\frac{1}{N_s}\left(\frac{v_r\, r}{k_{ps}} + r_s\right) - \alpha_2\right] \Delta\, C_S}{\frac{h}{N_s} k_{sed} (1 - f_p) - \frac{\left[\frac{1}{N_s}\left(\frac{v_r p}{k_{ps}} + r_s\right) - \alpha_2\right]\left[k_{out} + k_{sed} (1 - f_p) - \alpha_1\right]}{\frac{v_r p}{h\, k_{ps}}}}$$

$$A_{12} = \left[\frac{1}{N_s}\left(\frac{v_r p}{k_{ps}} + r_s\right) - \alpha_2\right] \frac{\frac{v_r p}{k\, k_{ps}} \Delta\, C_S - \left[k_{out} + k_{sed} (1 - f_p) - \alpha_1\right] \Delta C_L}{\frac{v_r p}{k_{ps}} \cdot \frac{k_{sed} (1 - f_p)}{N_s} - \left[\frac{1}{N_s}\left(\frac{v_r p}{k_{ps}} + r_s\right) - \alpha_2\right]\left[k_{out} + k_{sed} (1 - f_p) - \alpha_1\right]}$$

$$A_{21} = \frac{\frac{h}{N_s} k_{sed} (1 - f_p)\, \Delta C_L - \left[\frac{1}{N_s}\left(\frac{v_r p}{k_{ps}} + r_s\right) - \alpha_2\right] \Delta C_S}{\frac{\frac{v_p}{k_{ps}} \cdot \frac{1}{N_s} k_{sed} (1 - f_p)}{k_{out} + k_{sed} (1 - f_p) - \alpha_1} - \left[\frac{1}{N_s}\left(\frac{v_r p}{k_{ps}} + r_s\right) - \alpha_2\right]}$$

$$A_{22} = \frac{\frac{v_r p}{h\,k_{ps}}\,\Delta C_S - \left[k_{out} + k_{sed}\,(1-f_p) - \alpha_1\right]\Delta C_L}{\frac{v_r p}{h\,k_{ps}} - \frac{\left[\frac{I}{N_s}\left(\frac{v_r p}{k_{ps}} + r_s\right) - \alpha_2\right]\left[k_{out} + k_{sed}\,(1-f_p) - \alpha_1\right]}{\frac{h}{N_s}\,k_{sed}\,(1-f_p)}}$$

$$N_s = h_s\left[\frac{r}{k_{ps}} + (1-p)\,\rho_s\right]$$

Appendix II

Coefficients A_{ij} of Eq.(1) in Ch. 2 for the System Epilimnion-Hypolimnion

$$A_{11} = \frac{N_{11}}{D_{11}}$$

$$N_{11} = \left[k_{mix,h} f_{pe} + k_{sed,h}\,(1-f_{pe})\right] \Delta C_{et} - \left[k_{mix,h}\, f_{ph} + k_{sed,h}\,(1-f_{ph}) - \alpha_2\right] \Delta C_{ht}$$

$$D_{11} = k_{mix,h} f_{pe} + k_{sed,h}\,(1-f_{pe}) - \frac{[k_{mix,h}\, f_{ph} + k_{sed,h}\,(1-f_{ph}) - \alpha_2]\,[k_{out,e} + k_{mix,e} f_{pe} + k_{sed,e}\,(1-f_{pe}) - \alpha_1]}{k_{mix,e} f_{pe}}$$

$$A_{12} = \frac{N_{12}}{D_{12}}$$

$$N_{12} = \left[k_{mix,h} f_{ph} + k_{sed,h}\,(1-f_{ph})\,\alpha_2\right] \cdot \left\{k_{mix,e} f_{ph}\, \Delta C_{ht} - \left[k_{out,e} + k_{mix,e} f_{pe} + k_{sed,e}\,(1-f_{pe}) - \alpha_1\right] \Delta C_{et}\right\}$$

$$D_{12} = k_{mix,e} f_{ph} \left[k_{mix,h} f_{pe} + k_{sed,h}(1-f_{pe})\right] - \left[k_{mix,h} f_{ph} + k_{sed,h}(1-f_{ph}) - \alpha_2\right] \cdot \left[k_{out,e} + k_{mix,e} f_{pe} + k_{sed,e}(1-f_{pe}) - \alpha_1\right]$$

$$A_{21} = \frac{[k_{mix,h}f_{pe} + k_{sed,h}\,(1-f_{pe})]\,\Delta C_{et} - [k_{mix,h}f_{ph} + k_{sed,h}(1-f_{ph}) - \alpha_2]\,\Delta C_{ht}}{\dfrac{k_{mix,e}f_{ph}[k_{mix,h}f_{pe} + k_{sed,h}\,(1-f_{pe})]}{[k_{out,e} + k_{mix,e}f_{pe} + k_{sed,e}\,(1-f_{pe}) - \alpha_1]} - [k_{mix,h}f_{ph} + k_{sed,h}\,(1-f_{ph}) - \alpha_2]}$$

$$A_{22} = \frac{k_{mix,e}f_{ph}\,\Delta C_{ht} - [k_{out,e} + k_{mix,e}f_{pe} + k_{sed,e}\,(1-f_{pe}) - \alpha_1]\,\Delta C_{ht}}{k_{mix,e}f_{ph} - \dfrac{[k_{mix,h}f_{ph} + k_{sed,h}\,(1-f_{ph}) - \alpha_2]\,[k_{out,e} + k_{mix,e}f_{pe} + k_{sed,e}\,(1-f_{pe}) - \alpha_1]}{k_{mix,h}f_{pe} + k_{sed,h}\,(1-f_{pe})}}$$

Part III

INTERPRETATIVE AND PREDICTIVE MODELLING OF SUBALPINE LAKES

1. The Phosphorus Cycle in Lake Lugano

1.1 The eutrophication of Lake Lugano

Phytoplankton production and hypolimnetic oxygen consumption must be considered, to correctly estimate the trophic evolution of Lake Lugano during the last decades.

If the data of the period 1944-1958 are compared to those of the period 1972-1982, the phytoplankton density in the upper layers of the lake during summer appears to have risen from less than 1000 up to 200.000 cells cm^{-3}. Since the end of 1950's the blue-green alga Oscillatoria rubescens, which is tipical of eutrophicated water bodies, has become dominant in summer and autumn blooms in Lake Lugano.

The very bad oxygenation of the hypolimnetic waters in this lake is synthetically expressed by the constant decrease of the dissolved oxygen concentration at the end of the summer stagnation (average decrease 0.16 g m^{-3} yr^{-1} between 1946 and 1981). In the meromictic basin of Lugano-Gandria (maximum depth > 300 m) water with dissolved oxygen concentration greater than 2 g m–3 has never reached depths greater than 100 m between first 1970's and first 1980's.

Correspondingly, the total phosphorus loading per unit lake surface area that enters Lake Lugano from point easily interceptable sources has increased up to about twice the value which is considered as compatible with the multiple use of water, which is 1.5 g P m^{-2}. Average spring overturn concentration of dissolved phosphorus has increased 11 mg m^{-3} per year between 1964 and 1974.

The evolution of Lake Lugano towards heavy eutrophication has been favoured by its morpho-hydrological characteristics (tab.I). As a matter of fact, on the basis of the ratio mean depth/water mean residence time (8.7 m yr^{-1}) - with the phosphorus load as indicated above - the lake was included among the most eutrophicated water bodies (see Part I, ch.2), whose study was recommended by the O.C.D.E. Cooperative Programme (Alpine Project, 1972).

After 1977 two waste water treatment plants (Lugano and Mendrisio) have operated, to which about 80% of the resident population is connected (fig.1). The treated sewage from the town of Lugano is discharged into the river Vedeggio, the main tributary of the Agno bay, at the end of the chain of bays which constitute the South-western holomictic basin of the lake (see the bathymetric map of the Agno bay in fig.2).

Tab. I Morpho-hydrological data of Lake Lugano

Mean level (m a.s.l.)	271
Mean watershed altitude (m)	786
Surface area (km^2)	48,9
Watershed surface area (km^2)	615
Maximum depth (m)	322
Mean depth (m)	130
Volume (km^3)	6.5
Average outflow rate ($m^3\ s^{-1}$)	25.2
Estimated effective mean water residence time (yr)	15

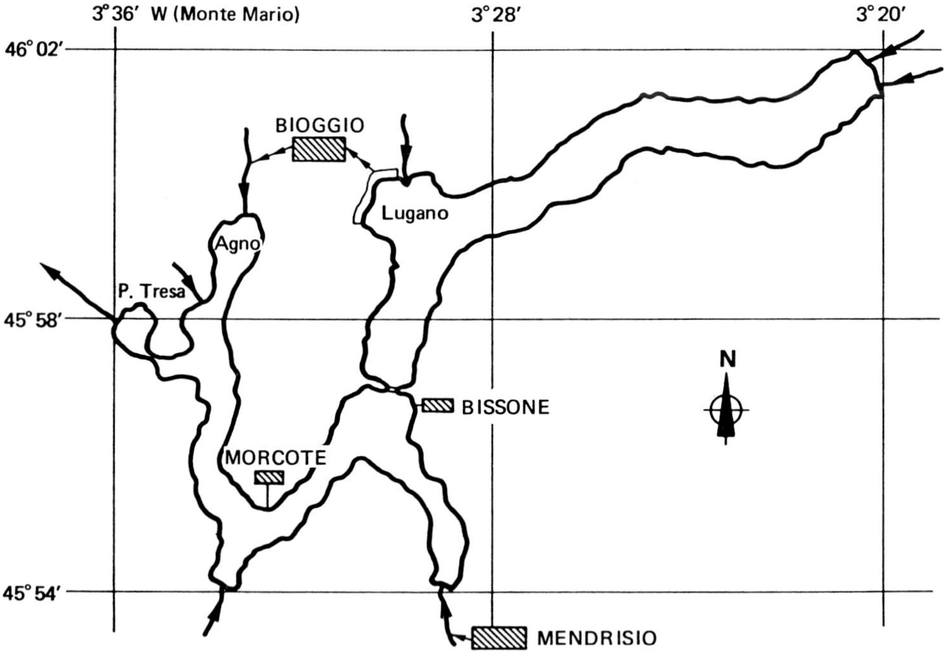

Fig. 1 Map of Lake Lugano with the location of the waste water treatment plants

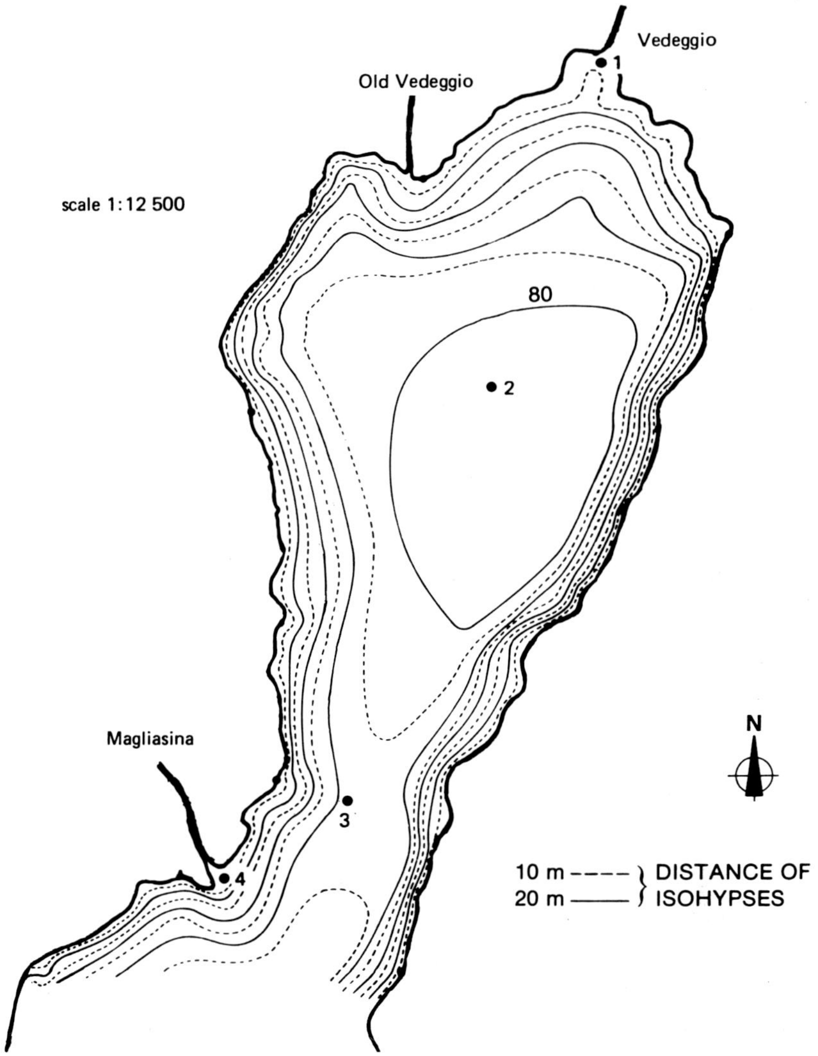

Fig. 2 Bathymetric map of the Agno bay. Numbers indicate water sampling stations.

For this reason, ecologic research in the Water Quality project of the J.R.C. Programme "Protection of the Environment" has been focussed on this bay from 1977 to 1981.

Two points were considered: the assessment of the internal phosphorus loading and the role of the dissolved organic component in the phosphorus cycle, both aspects which had been put in evidence by the implementation of the waste water treatment plant. The results are presented in section 1.2. After 1981 the control of the trophic evolution of Lake Lugano has been taken up by the Swiss Authority. A slight improvement of the situation described above, as regards both algal biomass and dissolved oxygen concentration, was observed towards the end of 1980's.

1.2 Interpretative modelling of the eutrophication process in Lake Lugano

The final version of the mathematical model, which has been developed during the project to describe the evolution of the eutrophication process in Lake Lugano, is a vertical box interaction model with three regions:

1) epilimnion;
2) hypolimnion;
3) bottom layer

The ecosystem constituents whose evolution is analysed in the model are:

1) phytoplankton (average algal population)
2) algal (intracellular) phosphorus
3) inorganic dissolved phosphorus
4) organic particulate phosphorus
5) organic dissolved phosphorus
6) dissolved oxygen

The interactions among the various constituents (fig. 3) describe the epilimnetic phosphorus cycle, as well as the "hypolimnetic" one going through the bottom sediments, phosphorus being considered as the only limiting nutrient (Part. I, ch. 1).

They are characterized by the following quantities (Part I, chs. 7,8):

1) Growth coefficient of the average algal population, assumed to be dependent on water temperature T, light intensity I, intracellular phosphorus concentration Q (mg P per mg biomass d.w.) as follows:

$$\mu = a \exp(bT) \frac{I(t)}{k_I + I(t)} \cdot \frac{Q(t) - Q_{min}}{Q(t)} \qquad (1)$$

   where k_I is the half-saturation constant for light dependence and Q_{min} the minimum internal phosphorus concentration.

2) Daily rate of transfer from algal to organic particulate phosphorus, taking into account all the losses of living phytoplankton biomass (mortality, settling, grazing).

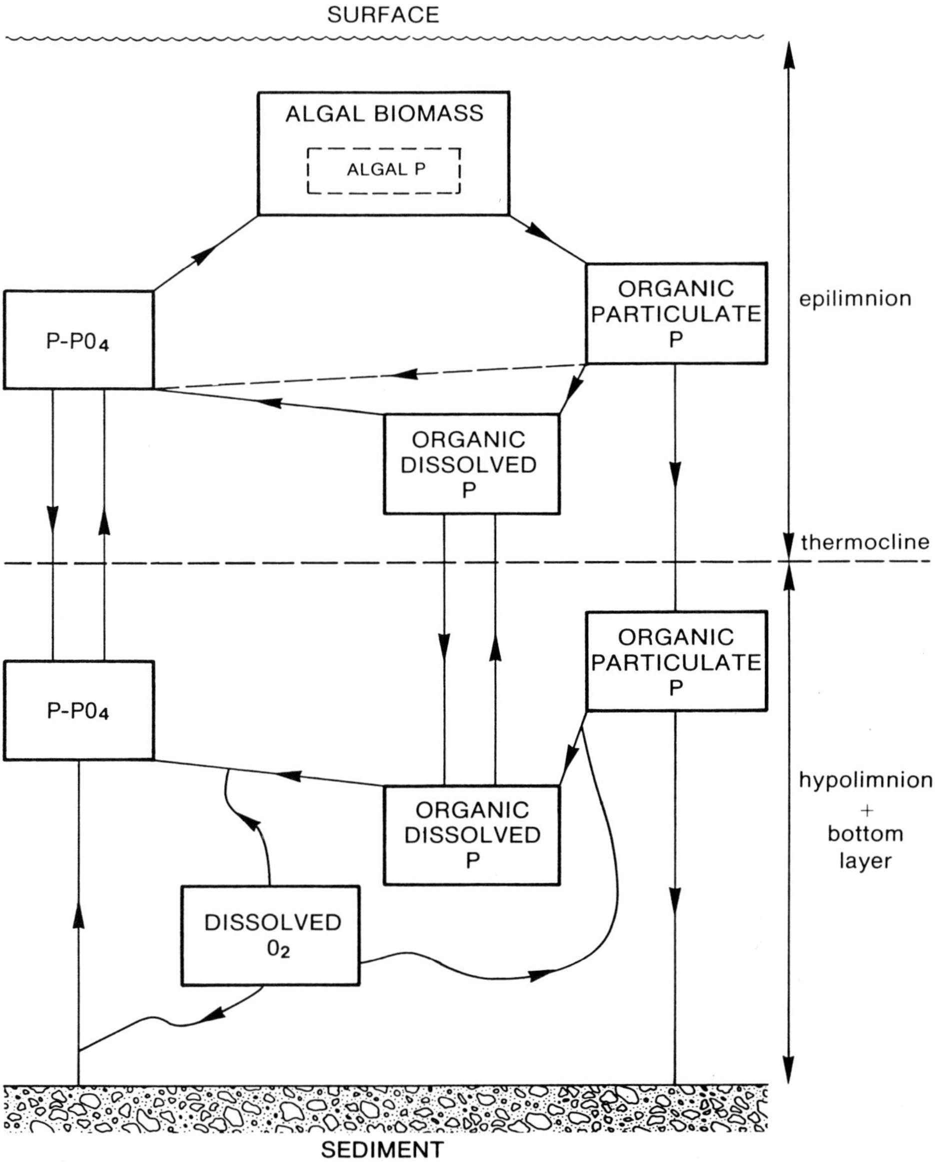

Fig. 3 Interactions among the various phosphorus forms considered in the lake eutrophication model

3) Settling velocity of algae and organic particulate.

4) Daily rate of transfer from organic particulate to organic dissolved phosphorus.

5) Daily rate of transfer from organic to inorganic dissolved phosphorus ($P\text{-}PO_4$).
6) Uptake rate of $P\text{-}PO_4$ by phytoplankton, assumed as dependent on external $P\text{-}PO_4$ concentration P_{id} and intracellular phosphorus concentration Q in the following way:

$$u = u_{max} \frac{P_{id}}{k_p + P_{id}} \frac{Q_{max} - Q}{Q_{max} - Q_{min}} \qquad (2)$$

where u_{max}, Q_{max} are the maximum values of u, Q and k_p is the half-saturation constant for external concentration dependence.

7) $P\text{-}PO_4$ flux from bottom sediments during a period of Δt days after t_o, when anoxic conditions begin at the water sediment interface, with time dependence assumed as sinusoidal:

$$\varphi(t) = \varphi_o \sin \frac{\pi}{\Delta t} (t - t_o) \qquad (3)$$

The main purpose of the model is to identify a set of parameters, which characterize the phosphorus cycle in the Agno bay during summer stagnation (morpho-hydrological data of the bay are given in tab. II).

Tab. II Morpho-hydrological data of the Agno bay

Surface area (km^2)	2.1
Watershed surface area (km^2)	102.7
Maximun depth (m)	87
Mean depth (m)	57
Volume (km^3)	0.121
Average outflow rate ($m^3\ s^{-1}$)	4
Estimated effective mean water residence time (yr)	2.2

Leaving to a further stage of research the quantitative description of the successive algal blooms taking place in the bay between April and October, the evolution of the phytoplankton concentration (mg algal biomass d.w. m^{-3} vs time) is assumed as known from experimental data (tab. III)

Tab. III Epilimnetic (0 - 10 m) average concentration of phytoplankton biomass (g d.w. m^{-3})

December 1980	0.467
January 1981	0.284
February	0.239
March	0.252
April	0.816
May	0.360
June	0.454
July	0.393
August	0.568
September	0.291
October	0.799
November	0.693
December	0.486
January 1982	0.163
Average July - October 1981	0.513 + 0.222

In other words, the model aims to identify a phosphorus cycle compatible with the observed algal biomass summer level.

The procedure is based on a comparison between calculated and measured average concentrations of the various phosphorus forms in epilimnion (0 – 10 m), hypolimnion (10 – 70 m) and bottom layer (70 – 87 m). It refers to the summer 1981, during which concentration profiles were measured in the centre of the bay (point of maximum depth ~ 87 m) approximately every fortnight.

Experimental results are summarized in tab. IV in terms of average concentrations for the period July-October 1981 and mean deviations.

These are relatively high in the epilimnion for all the phosphorus forms and along the whole water column for the organic dissolved component.

As a matter of fact the epilimnetic phenomena appear as more complex, and the experimental framework is less clearly defined in the epilimnion than in the hypolimnion, a better definition asking for more frequent sampling, which was prevented by logistic difficulties.

Tab. IV Different fractions of phosphorus in the water column of the Agno basin (summer 1981)

($mg\ m^{-3}$)	**$P\text{-}PO_4$**	**$P_{org\ part}$**	**$P_{org\ diss}$**
0 - 10 m	7.4 ± 4	24.4 ± 6.3*	81.5 ± 43.2
10 - 70 m	101.4 ± 12	2.4 ± 0.7	36.3 ± 22
78 m	141.1 ± 11	11.1 ± 3.6	31.3 ± 24.2

* including algal phosphorus (estimated concentration 12.2 ± 3 mg m–3)

Therefore the model gives up interpreting the level fluctuations of the single phosphorus forms and only relys upon their average values during summer. The assessment of the transfer coefficients is considered as a part of a iterative procedure, in order to compare different possible interpretative schemes, and for this reason it is presented as a *correlation* between theory and experiments more properly than a simple model calibration.

The results must be considered as indications for an acceptable average summer value of the coefficients concerned which, particularly as regards the organic phosphorus components, could hardly be drawn from a direct inspection of their experimental mass balances in the epilimnion.

1.2.1 The "hypolimnetic" phosphorus mass balance

It had been previously demonstrated in the case of the Ponte Tresa basin of Lake Lugano that the prevailing term of the hypolimnetic phosphorus mass balance related to the summer anoxic period is the $P\text{-}P0_4$ release flux from the sediments. In the Agno bay, however, only the deepest layers exhibit a sharp increase of the $P\text{-}PO_4$ concentration during summer with respect to the spring values. For this reason, in order to assess the $P\text{-}P0_4$ flux from the sediments of this bay, a bottom layer (70 – 87 m, 8.1 . $10^6\ m^3$, area of the projected bottom surface 0.9225 . $10^6\ m^2$) is considered as separated from the hypolimnion (10 – 70 m, 92.4 . $10^6\ m^3$). This is justified by the fact that only below 70 m depth approximately anoxic conditions exist in this bay during summer.

Considering as negligible the $P\text{-}PO_4$ flux from the sediments in aerobic conditions, the following coefficients are assumed as known:

- the mixing coefficient at the depth of the thermocline $m_{th} = 0.1\ md^{-1}$, derived from measured temperature profiles;
- the mineralization rate of organic phosphorus in the hypolimnion $k_{min}^{hyp} = 0.002\ d^{-1}$, derived from previous experimental investigation in the bay.

Then the balance equation for inorganic dissolved phosphorus P_{id} *in the bottom layer*

$$V\ d/dt\ P_{id} = S\ \varphi\ (t) + Sm_{bl}\ (P_{id}^{hyp} - P_{id}) + \ldots \qquad (4)$$

contains two unknown parameters:

φ_o	=	peak value of the P-PO_4 flux from the sediments $\varphi(t)$ (see eq. (3));
m_{bl}	=	mixing coefficient at the interface between bottom layer and hypolimnion (the other symbols are obvious).

The procedure starts with an initial estimate of m_{bl} (which can be supported by auxiliary one-dimensional calculations of P-PO_4 diffusion in the hypolimnion) and consists of identifying the value of φ_o which gives a satisfactory agreement of the calculated $P_{id}(t)$ curve with the experimental results.

Introducing the obtained $P_{id}(t)$ and the estimated m_{bl} in the balance equation for P-PO_4 *in the hypolimnion* and comparing the $P_{id}^{hyp}(t)$ curve with the experimental one, allow to correct the initial estimate on m_{bl} and consequently the value of φ_o.

The correlation applied to the Agno bay gives the following results:

$$m_{bl} = 5m_{th} = 0.5 \text{ md}^{-1}$$

$$\varphi_o = 0.025 \text{ g P m}^{-2} \text{ d}^{-1}$$

(see fig. 4). A variation of the parameter m_{bl} would imply a variation of φ_o in the same direction, and this would worsen the agreement with the experimental results as for the P-$P0_4$ level in the hypolimnion.

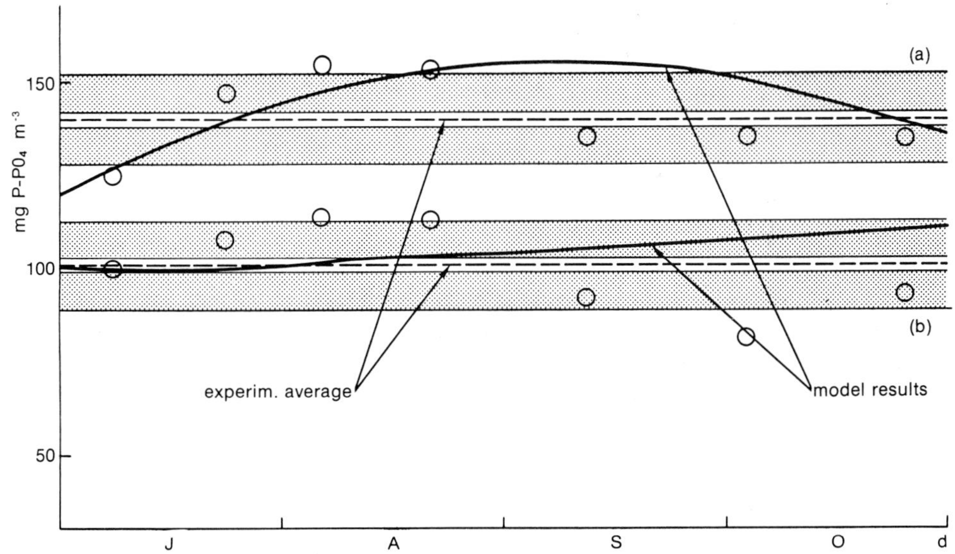

Fig. 4 Average P - PO_4 concentration in the bottom layer (a) and in the hypolimnion (b) of the Agno bay.
Solid line: model results.
Shaded area: experimental average concentration ± standard deviation.

1.2.2 The epilimnetic phosphorus mass balance

The experimental results presented in tab. IV need some additional information and comments.

The epilimnetic average concentration of organic particulate phosphorus was obtained from the measured particulate phosphorus concentration (35. ± 9. mg P m^{-3}) by applying a factor 0.7 as suggested by the analysis of particulate material collected by means of sedimentation traps at –15 m.

Particulate organic phosphorus was considered 50% algal and 50% non-algal as an average estimate for the 0 - 10 m water column (in reality the fraction of the algal phosphorus can reach 100% of the organic particulate in the upper meters).

The average epilimnetic phosphorus uptake rate by phytoplankton was drawn form measurements of C assimilation in the euphotic zone (0 - 3 m). The average summer value of primary productivity was evaluated to be 2650 mg C m^{-2} d^{-1} for an average standing crop of 1856 mg C m^{-2}, the growth coefficient assuming then the value 1.43 d^{-1} (photosynthetic activity below 3 m depth was neglected).

In the euphotic zone algal P was assumed to be between 50% and 80% of the particulate fraction. This corresponds to a C : P ratio between 40 and 25 : 1. Then, the above indicated average level of C assimilation and the assumption of steady state as regards the C : P ratio inside the biomass allow the estimate of the lower and upper limits of the average P uptake rate during summer as 66.3 and 106 mg P m^{-2} d^{-1}, respectively.

Referring to *the epilimnion (0 - 10 m)*, a measured average concentration of 7.4 mg P-PO_4 m^{-3} and an estimated average concentration of 12.2 mg algal P m^{-3}, with an average uptake rate of 8.6 mg P m^{-3} d^{-1}, allow the estimate of the mean residence time of phosphorus in the compartments of dissolved inorganic and algal P as 0.9 and 1.4 d respectively.

It is worth pointing out that, even considering the large fluctuations, the amount of organic dissolved phosporus is sharply greater than that of particulate within the epilimnetic waters. The epilimnetic organic dissolved phosphorus level is sustained both by lytic processes concerning the organic particulate and by the external loading originated from the effluents of the depuration plant. Mineralization processes affecting the dissolved fraction of the organic phosphorus pool can be considered as rather slow on the average.

Then, two assuptions are suggested for the role of the organic dissolved component in the epilimnetic phosphorus cycle:

a) all regenerated P-PO_4 comes from the organic dissolved P pool (scheme 1 of fig. 5);

b) there is a more efficient "direct" mineralization process acting on the organic particulate P and a less efficient one on organic dissolved P, at the same slow rate (some 0.1% d^{-1}) constantly observed in the hypolimnetic water column of the basin (scheme 2 of fig. 5).

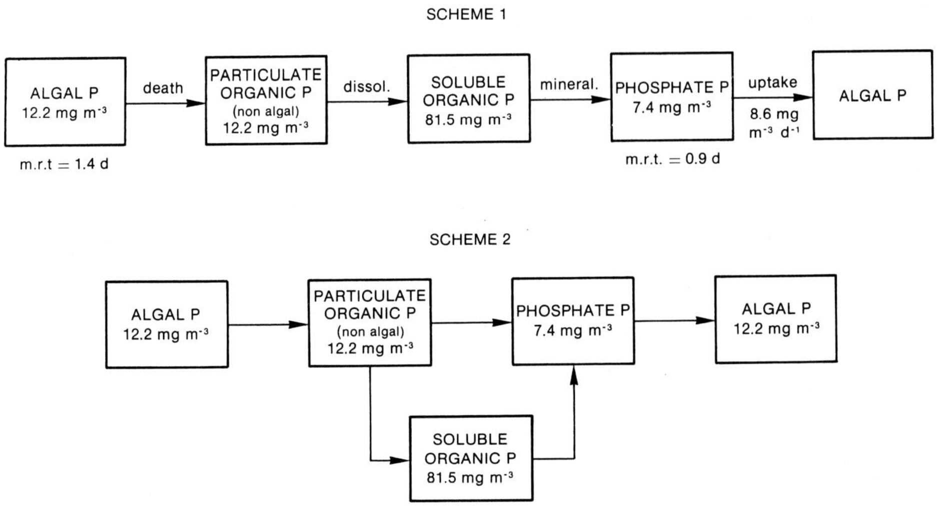

Fig. 5 Schematic diagrams of the phosphorus cycle in the epilimnion of the Agno bay during summer stagnation.
P concentrations and mean residence times (m.r.t.) are based on experimental evaluations.

The question is whether, and for which values of the transfer coefficients, the two assumptions are compatible with the observed concentrations of the phosphorus forms in the bay.

a) The parameters which characterize the phosphorus cycle, if all the regenerated P-P04 is assumed to come from the organic dissolved phosphorls pool, are the following:
k_1 = daily rate of transfer from algal to organic particulate (non-algal) P;
k_2 = daily rate of transfer from organic particulate to organic dissolved P;
k_3 = daily rate of transfer from organic dissolved to $P\text{-}P0_4$ (inorganic dissolved);
k_4 = maximum uptake rate of $P\text{-}P0_4$ by phytoplankton.
The parameters k_1, k_2 enter the balance equation for organic particulate (non-algal) phosphorus P_{op}; k_2, k_3 enter the equation for organic dissolved phosphorus P_{od}; k_3, k_4 enter the equation for $P\text{-}P0_4$ (P_{id}), in the following way:

$$d/dt\, P_{op} = k_1 B\,(t)\, Q\,(t) + q_{in}/V\, \rho_{ext}^{op} - (k_2 + v_s/h)\, P_{op}$$

$$d/dt\, P_{od} = k_2 P_{op} + q_{in}/V\, \rho_{ext}^{od} - (k_3 + q_{out}/V)\, P_{od} + m_{th}\, S_{th}/V\, (P_{od}^{hyp} - P_{od})$$

$$d/dt\, P_{id} = k_3 P_{od} + q_{in}/V\, \rho_{ext}^{id} - k_4 P_{id}/(k_p + P_{id}) \cdot (Q_{max} - Q)/(Q_{max} - Q_{min})\, B\,(t)$$

$$- q_{out}/V\, P_{id} + m_{th}\, S_{th}/V\, (P_{id}^{hyp} - P_{id}) \qquad (5)$$

where B(t) is the phytoplankton biomass concentration (d.w.) drawn from measurements (tab. III); V, h are respectively volume and thickness of the epilim-

nion; S_{th} is the thermocline surface area. k_p, Q_{max}, Q_{min}, m_{th}, v_s (settling velocity of the organic detritus) are considered as known, the adopted values being based on experimental investigations or taken from literature. Inflow-outflow rates $q_{in} = q_{out}$ and phosphorus concentrations in the tributaries ρ_{ext} have been monitored along the year.
With these premises the system of non-linear differential equations (5) is numerically solved (STIFF method) in order to identify a set of values k_1, k_2, k_3, k_4, for which the calculated functions P_{op} (t), P_{od} (t) P_{id} (t) are in satisfactory agreement with the experimental summer average value of the related phosphorus concentrations.
It is obviously necessary to start with an initial estimate for k_1, and to utilize the first equation (5) for the empirical identification of k_2, after which the obtained value of k_2 is introduced in the second equation and k_3 is identified, and so on (empirical identification means to vary the concerned parameter until a "satisfactory" agreement of the equation solution with the experiments is obtained).
If an expression of the type

$$k_1 = k_{10} + k_{11} B(t) (T - T_o) \tag{6}$$

is adopted, the initial estimate concerns the constants k_{10}, k_{11}; k_{10} can be chosen on the basis of the assumption of steady state for the phytoplankton during winter water omothermy ($T = T_o$), i.e. on thè basis of the value of the algal growth coefficient for $T = T_o$.
An experimental indication for the choice of k_{11} is given by the above estimated mean residence time of phosphorus in the algal pool (~ 1.4 d): k_{11} has to be chosen in such a way that the average summer value of k_1 is compatible with this mean residence time.
A final check of the procedure is provided by the estimated average phosphorus uptake rate by the phytoplankton (0.013 ÷ 0.021 d^{-1} if related to the average d.w. phytoplankton concentration).
As a matter of fact the identified k_4 value allows to calculate the time evolution u(t) of the P uptake rate, to be compared with the above indicated interval of acceptable values.
Starting with the values $k_{10} = 0.35\ d^{-1}$, $k_{11} = 0.03\ (mg\ m^{-3})^{-1}\ °C^{-1}\ d^{-1}$, which give a summer average value of $k_1 \sim 0.7\ d^{-1}$ as suggested by the experimental estimates, the comparison with the experimental summer average values for organic particulate P, organic dissolved P and inorganic dissolved P (figs. 6, 7, 8, solid lines) indicates as compatible with the experiments the following value of the transfer coefficients:

$$k_2 = 0.5\ d^{-1};\ k_3 = 0.07\ d^{-1};\ k_4 = 0.2\ d^{-1}.$$

In fig. 9 the time evolution of the P uptake rate u (t) (mg P per mg algal biomass d.w. per day), as calculated by the model on the basis of the parameter values suggested by the correlation is presented: the (solid) curve is well located between the experimental limits.

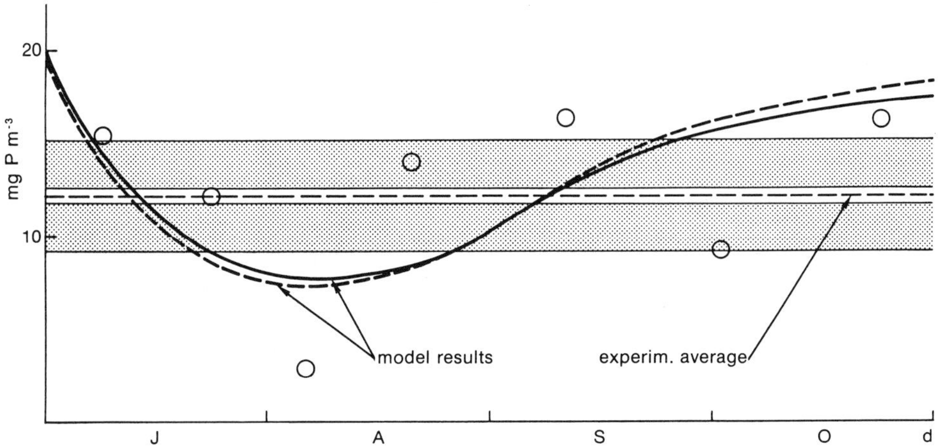

Fig. 6 Average organic particulate (non-algal) P concentration in the epilimnion of the Agno bay
Solid line: model results following scheme 1 of fig. 5.
Dashed line: model results following scheme 2 of fig. 5.
Shaded area: experimental average concentration ± standard deviation.
○ Experimental values.

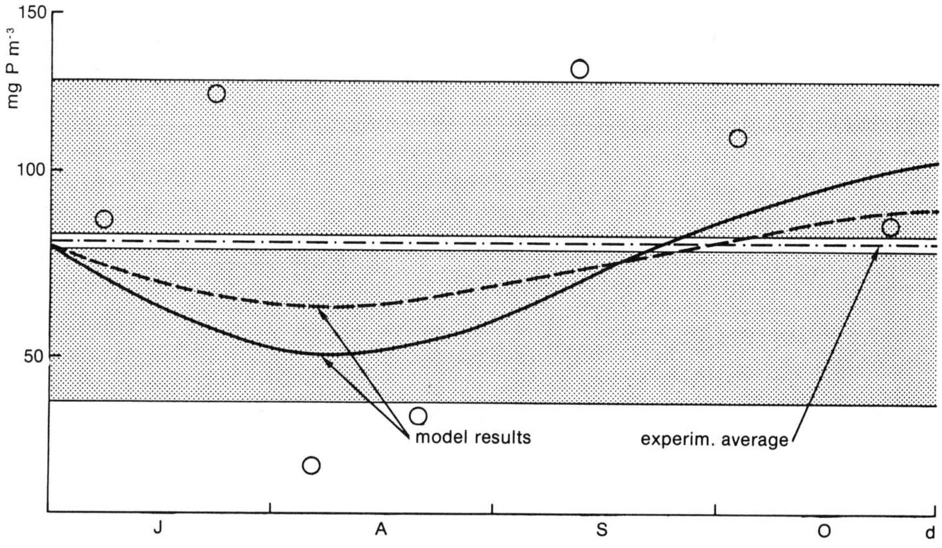

Fig. 7 Average organic dissolved P concentration in the epilimnion of the Agno bay.
Solid line: model results following scheme 1 of fig. 5.
Dashed line: model results following scheme 2 of fig. 5.
Shared area: experimental average concentration ± standard deviation.
○ Experimental values.

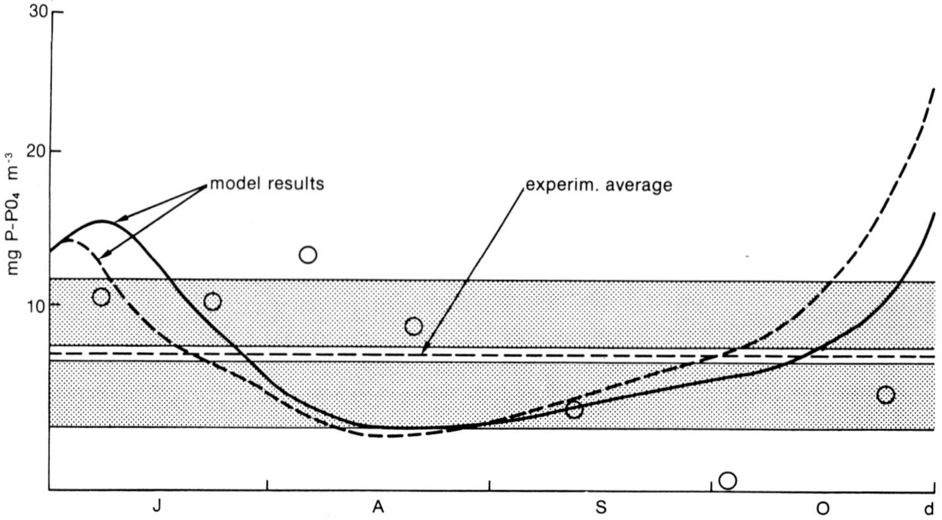

Fig. 8 Average P-PO4 concentration in the epilimnion of the Agno bay.
Solid line: model results following scheme 1 of fig. 5.
Dashed line: model results following scheme 2 of fig. 5.
Shaded area: experimental average concentration ± standard deviation.
O Experimental values.

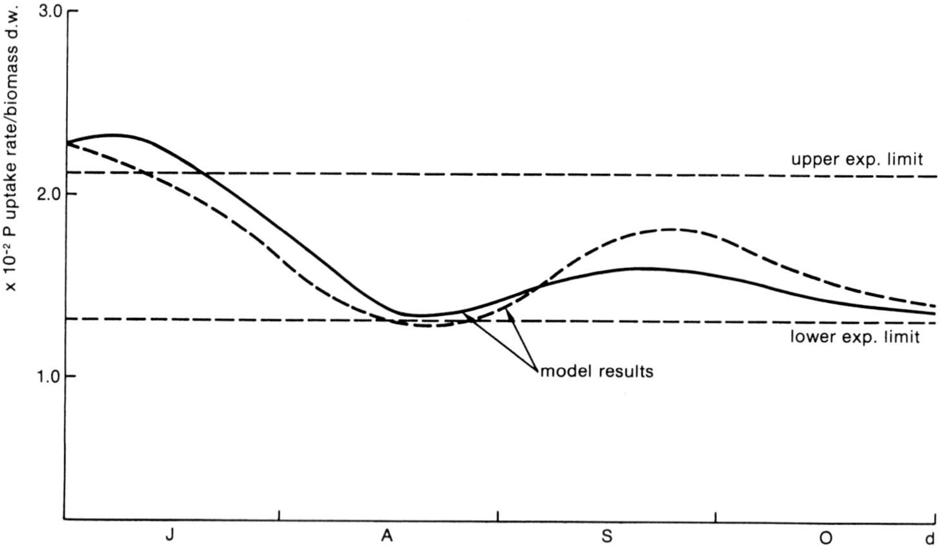

Fig. 9 Average P uptake rate referred to algal biomass d.w. in the epilimnion of the Agno bay
Solid line: model results following scheme 1 of fig. 5.
Dashed line: model results following scheme 2 of fig. 5.
- - - - - upper and lower experimental limits.

b) If the P-P0$_4$ regeneration process is assumed to concern mainly organic particulate and only at a much slower rate organic dissolved phosphorus, the parameter k_3 is defined as the transfer coefficient from organic particulate to inorganic dissolved phosphorus.
The mineralization rate of organic dissolved phosphorus in the epilimnion is assumed to be known, with the same value as in the hypolimnion (0.002 d^{-1}).
In this case the transfer coefficients enter the balance equations in the following way:

$$\begin{aligned} d/dt\, P_{op} &= k_1\, B(t)\, Q(t) - (k_2 + k_3)\, P_{op} + \ldots \\ d/dt\, P_{od} &= k_2\, P_{op} + \ldots \\ d/dt\, P_{id} &= k_3\, P_{op} - k_4 \frac{P_{id}}{k_p + P_{id}} \cdot \frac{Q_{max} - Q(t)}{Q_{max} - Q_{min}} \cdot B(t) + \ldots \end{aligned} \qquad (7)$$

the other terms being the same as in eqs. 5.
The procedure consists of identifying the total "loss" coefficient of organic particulate phosphorus ($k_2 + k_3$) by means of the first equation of the system (7), and the "lysis" coefficient k_2 by means of the second equation, then of deducing k_3 as a difference, and identifying k_4 by means of the third equation.
All the above accessory considerations could be repeated and, for the same values of k_{10}, k_{11}, the comparison with the experiments (dashed curves in figs. 6, 7, 8) gives the following values of the transfer coefficients:

$$k_2 = 0.1\, d^{-1};\; k_3 = 0.4\, d^{-1};\; k_4 = 0.2\, d^{-1}$$

Furthermore, the calculated P-PO$_4$ uptake rate vs time u(t) agrees with the experimental estimates in the same satisfactory way as in the previously considered scheme (fig. 9, dashed curve).
It might be concluded that the measured summer average concentrations are compatible with both the considered schemes for the epilimnetic phosphorus cycle, provided the transfer coefficients be given appropriate values.
The assessment of the transfer coefflcients among the various phosphorus compartments suggests the following comments on the epilimnetic phosphorus residence times.
It must be distinguished between residence time of phosphorus in the epilimnion and turnover time as the sum of the characteristic times of the epilimnetic phosphorus transfer processes.
In steady state conditions the phosphorus residence time in the epilimnion is given by

$$\tau_{epi} = \frac{P_d + P_p}{P_d \cdot q_{out}/V_{epi} + P_p \cdot v_s/h} \qquad (8)$$

where P_d, P_p are the (summer average) epilimnetic concentrations of the total dissolved and total particulate phosphorus; V_{epi}, h volume and thickness of the epilimnion; S_{th} the surface area of the thermocline, q_{out} the outflow rate and v_s the settling velocity of the particulate matter.

In the Agno bay it has been found $\tau_{epi} \sim 35$ d for summer 1981.

The characteristic times of the epilimnetic phosphorus transfer processes assessed in this paper sum up to ~ 19 d (case a) and ~ 5 d (case b) as the upper and lower limits of the epilimnetic phosphorus turnover time. So this time resulted as significantly lower than the residence time in the epilimnion, and this is consistent with the observed high incidence of phosphorus into the productivity of the bay.

A concluding comparison between calculated curves as resulting from the correlation and experimental values of the total phosphorus concentration in the epilimnion of the Agno bay is shown in fig. 10.

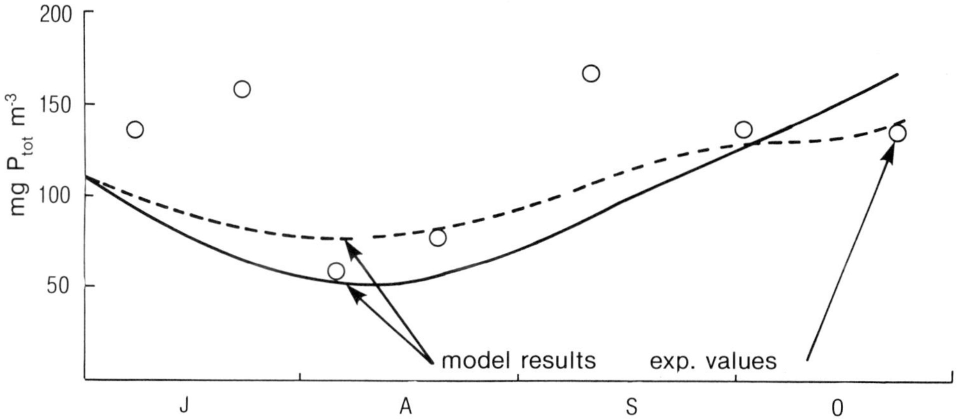

Fig. 10 Average total phosphorus concentration in the epilimnion of the Agno bay.
Solid line: model results following scheme 1 of fig. 5.
Dashed line: model results dollowing scheme 2 of fig. 5.
O Experimental values.

The measurements contain the contribution frcm particulate inorganic phosphorus, which is not taken into account by the model.

It must be observed that the concentrations resulting from the second sampling of July and the first sampling of September correspond to peak values which the model is unable to interprete on the basis of measured phosphorus loads and constant transfer rates. These peak values are due to the dissolved organic component, as it can be ascertained from the experimental points of fig. 7. Then it becomes evident that a detailed interpretation would require (besides more frequent sampling) to take into account other processes affecting this component.

Although it appears that only orthophospate P can be directly absorbed by algal cells, other forms may be available after hydrolysis by enzymes produced by algae themselves. Unfortunately, particulate and dissolved alkaline phosphatase activities were not determined in the Agno bay during summer 1981. At present is ignored what fraction of the dissolved organic phosphorus pool is available for phytoplankton assimilation. The nature of soluble organic phosphorus (unreactive P) is largely unknown, so as its role in the phosphorus economy of the epilimnetic layer. The presented correlation emphasizes that a phosphorus model must treat at least four phosphorus components: algal, organic particulate, organic dissolved and phosphate P.

Sources and suggestions for further reading

1. G. Premazzi and G. Rossi, Phosphorus Cycle in a Eutrophic Subalpine Lake Rep. EUR 9116 EN (1984)
2. G. Rossi, et al., Correlation of a lake eutrophication model to field experiments Ecol. Model, 34, 167-189 (1986)
   (Drawings were reprinted from this paper with permission of Elsevier Science Publishers B.V.)

2. Delay in Recovery of Lake Varese Caused by Internal Loading

2.1 The eutrophication of Lake Varese

Lake Varese was the first lake in Italy to experience problems of cultural eutrophication: this happened in the middle 1950's. The main problems identified in this lake are:

- dense algal blooms (Cyanophyceae dominance) from March to November (average summer density: 80. 10^6 cells cm^{-3})
- high oxygen consumption in hypolimnetic waters (estimated daily consumption rate: 0.05 g m^{-3} d^{-1} from March to June).

The bathymetric map of Lake Varese (fig.1) puts in evidence a shallow eastern bay (Schiranna) and a deeper north-western one (Gavirate), whose sediments have different properties and behaviour, while the sediment of a third smaller bay (Cassinetta) are considered as inactive.

Morpho-hydrological data of the lake are summarized in tab.I. Inflowing waters are

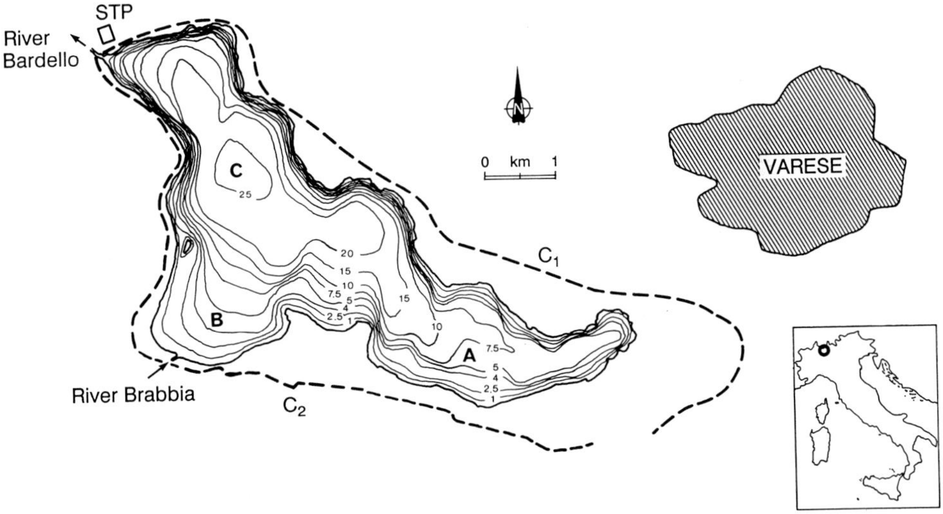

Fig. 1 Bathymetric map of Lake Varese (A Schiranna basin; B Cassinetta basin; C Gavirate basin). The two collectors (C1, C2) and the centralized sewage tratment plant (STP) are also indicated.

Tab. I Morpho-hydrological data of Lake Varese.

Mean level (m a.s.l.)	238
Surface area (km^2)	14.5
Watershed surface area (km^2)	111.5
Volume (km^3)	0.154
Maximum depth (m)	26
Mean depth (m)	11
Average outflow rate ($m^3\ s^{-1}$)	2.5
Estimated effective mean water residence time (yr)	2.3

subdivided among various small streams: the only permanent tributary is the Canale Brabbia (average inflow rate ~ 1 $m^3\ s^{-1}$), flowing from the shallow heavily eutrophicated Lake Comabbio. The emissary (river Bardello, average outflow rate ~ 2.5 $m^3\ s^{-1}$) is a tributary of Lake Maggiore.

The recovery objective was established by Regional Authorities with reference to the natural background phosphorus level of the lake (18 mg P m^{-3}), with an intermediate objective (acceptable level) 100% and a final objective 50% above it.
The planned restoration measures consist of sewage diversion and a waste water treatment plant, whose effluents flow into the emissary. The treatment plant was implemented at the beginning of 1987. The consequent reduction of external phosphorus loading was estimated from 45 to 25 t P yr^{-1}. Then release from sediments, which had been previously assessed as 70% of the external loading, became the major source of phosphorus for the lake.

2.2 Predictive one box modelling of Lake Varese recovery

The characteristic time of the transient phase after reduction of external P loading and the final in-lake phosphorus concentration can be approximately evaluated by means of a simple balance equation for total phosphorus concentration P_L, provided that net sedimentation (i.e. gross sedimentation minus release from sediments) can be assumed as constant (i.e. independent of phosphorus concentration) at the pre-diversion level.

The balance equation is as follows (Part I, ch.2):

$$V \frac{dP_L}{dt} = - q_{out}\ \beta P_L + q_{in}\ P_{ext} - S* \tag{1}$$

where:

V	=	lake volume;
q_{out}	=	outflow rate;
q_{in}	=	inflow rate;
P_{ext}	=	P concentration in inflowing waters;
S*	=	net sedimentation;
β	=	ratio between phosphorus concentration in the mixed surface layer and average phosphorus concentration in the water column.

The characteristic time of the evolution of P_L as described by (1) is:

$$\tau = \frac{V}{q_{out}\,\beta}$$

and the asymptotic ($t \longrightarrow \infty$) value of total phosphorus concentration is:

$$P_{L\infty} = \tau \frac{q_{in}\,P_{ext} - S*}{V} \qquad (2)$$

An indirect experimental verification of the assumption of constant net sedimentation can be obtained by means of total phosphorus mass balances, for instance between two successive water overturns (Δt days), because:

$$S*\,\Delta t = q_{in}\,P_{ext}\,\Delta t - q_{out}\,\overline{P}_L - V\Delta P_{L,} \qquad (3)$$

where ΔP_L is the variation of the concentration and $\overline{P}_L$ is the average phosphorus concentration in the effluent during the considered time interval.

In the case of Lake Varese this assumption has been satisfactorily verified for the years 1985-88, external P loading being reduced from ~ 45 t to ~ 25 t during 1987.
With an external P loading of about 25 t P, a net sedimentation S* = 17 t P and β ~ 0.7, we find:

$$\tau \approx 3 \text{ yr}$$
$$P_{L\infty} \approx 152 \text{ mg m}^{-3}$$

to be compared with the previous total P concentration of 350 mg m^{-3}, but still too high to be acceptable.

Obviously eq. (2) cannot be used to evaluate the external P loading compatible with an acceptable steady state concentration, because net sedimentation will tend to vary after a certain time (e.g. a decrease of gross sedimentation with constant or even increasing release flux from the sediments can be expected).

For this reason even the simplest model, aiming at a very "schematic" prediction of eutrophic lake recovery, must consider the system lake-sediment, i.e. it must take into account the dependence of gross sedimentation on the phosphorus concentration in lake, as well as the dependence of phosphorus release on the distribution of sedimentary phosphorus among the various compartments (organic, inorganic dissolved, inorganic adsorbed).

2.3 Modelling water-sediment interaction in Lake Varese

A lake-sediment interaction model such as presented in Part I, ch. 4, allows to predict Lake Varese recovery independently of the assumption of constant net sedimentation. Referring us to the system (6), the equation coefficients are evaluated on the basis of recent experimental studies on the lake, including analysis of P concentration profiles in sediment cores.

As for the *water column* (mean depth $\bar{z}$ = 11 m), the fraction particulate P/total P is taken = 0.15, and the settling velocity v_s = 0.2 m d^{-1} = 73 m yr^{-1}, as in the framework of an ecological model recently validated for lake Varese.
As for *sedimentary quantities*, average values are obtained by taking separately into account the "deep" bottom area (Gavirate basin, 8.6 km^2, mean depth 14.5 m) and the "shallow" one (Schiranna basin, 4 km^2, mean depth 5.5 m), and considering the sediments of the Cassinetta basin (1.9 km^2) as inactive (note that non uniformity of sediment properties, due to the presence of large shallow bays or coastal zones, could suggest the application of the model to these lake regions separately, introducing appropriate horizontal exchange terms in the respective balance equations).
The release "velocity" v_r is evaluated with reference to the average release flux φ = 7,3 mg P m^{-2} d^{-1} as experimentally assessed in 1987 (internal P loading 34 t yr^{-1}), and to the P concentration in pore water of the active sediment P_{int} ~ 2500 mg m^{-3}: the adopted value is v_r = 1.1 m yr^{-1}.
The most recent evaluations of the sedimentation rate (Cs-137 method) have given as average value for Lake Varese r_s = 1560 g m^{-2} yr^{-1}.
The active sediment layer results to have h_s = 0.09 m as average thickness ($V_s = h_s A_s$ = 1.134 x 10^6 m^3), average porosity p = 0.9, average density of solid sediment ρ_s = 2.4 x 10^6 g m^{-3} (total solid sediment mass of the sediment in the active layer M = 271 x 10^9 g).
The unexchangeable P concentration is given by the average concentration in the deepest layers of the sediment cores, i.e. $P_{ads,o}$ ~ 0.8 mg g^{-1} (fig. 2).
Laboratory resuspension experiments on sediment cores, sampled at various stations in different seasons, showed that the sorption sites of the sediment were already fully utilized in the active layer. The saturation concentration was estimated to range from 1.2 to 1.5 mg P g^{-1}.
The present concentration of dissolved P in pore water being P_{int} ~ 2500 mg m^{-3} as average value for the active layer, the saturation concentration $P_{int,s}$ can be estimated to range from 1500 to 2000 mg m^{-3}.

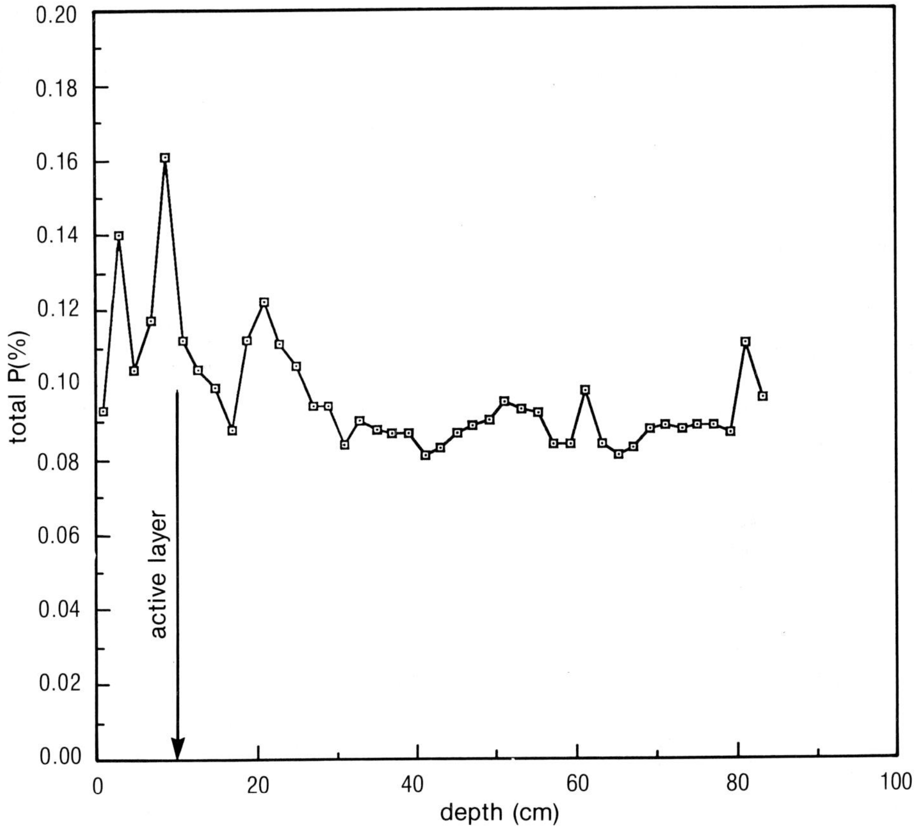

Fig. 2 Characteristic total P concentration in sediments of Lake Varese (deep area). Values are expressed as percentage of dry weight of sediment.

Hence the distribution constant a, which is expected to characterize the sediment dynamics in Lake Varese after the transition to unsaturated sediment, can be estimated to vary from 0.2 to 0.5 10^{-3} m^3 g^{-1}.

In the simplified scheme as described in Part I, ch.4 the recovery of a P-saturated sediment is not included. As a matter of fact the solutions of the system (6) are decreasing functions of time, even if $P_{ads}(0) = P_{ads,s}$.

On the contrary the recovery of a saturated sediment requires as a necessary condition that, in sedimenting solid particles, sites for P adsorption are available, i.e. a lake state characterized by:

$$\frac{v_s \pi P_L}{r_s} < P_{ads,s}$$

i.e.

$$P_L < P_L^* = \frac{r_s}{v_s \pi} P_{ads,s} \qquad (1)$$

Furthermore as long as $P_{int} > P_{int,s}$, dissolved P in pore water would saturate the fresh unsaturated sediment.

In the transition from saturated to unsaturated sediment the description of the sediment evolution by means of a single state variable is obviously impossible. If $P_{int}(0) > P_{int,s}$, the system of eq. (6) gives no information on the evolution of dissolved P in pore water towards $P_{int.s}$; moreover it underevaluates P release flux from the sediment, hence the time t* needed by P_L to reach P_L^*.

In order to estimate t*, the only way is to replace the system of eq. (6) with the following one:

$$\frac{dP_L}{dt} = -(k_{out} + k_{sed}) P_L + \frac{v_r p}{\bar{z}} P_{int} + k_{in} P_{ext}$$

$$\frac{dP_{int}}{dt} = \frac{v_s \pi P_L}{p\, h_s} - \frac{v_r}{h_s} P_{int} - \frac{r_s P_{ads,s}}{p\, h_s} \qquad (2)$$

The first equation is the same as in (6) with the release flux directly expressed in terms of P_{int}. The second one can be considered as a correct total P mass balance in the active sediment layer as long as P_{ads} = const = $P_{ads,s}$ (i.e. as long as $P_L > P_L^*$). At this time (t = t*) the sink term $- r_s P_{ads,s}/(ph_s)$ begins to decrease and the description of the sediment evolution would require a third equation.

For this reason, treating the system (2) with the standard analytical procedure for the solution of systems of linear differential equations with constant coefficients, would give incorrect results. A careful evaluation of t* would require a numerical solution of the system (2), but a rough estimation can be simply obtained by means of its time constant α_2.

This constant results to be significantly greater than the time constant of the system of eq. (6), so that a separate description of the transition to partially unsaturated sediment might be considered as concerning a minor detail in the framework of a long term prediction on the evolution of P in-lake concentration.

In the case of Lake Varese $\alpha_2 \sim 0.3\ yr^{-1}$ ($\alpha_2^{-1} \sim 3.3$ yr), the concentration P_L^* ranging from 170 to 210 mg m^{-3} depending on the adopted value of $P_{ads,s}$.

The additional delay-time before the description of the evolution of the system lake-sediment in terms of eqs. (6) becomes fully justified, depends on the intensity of mixing processes inside the active sediment layer, i.e. it is expected to be longer in the shallow region of the lake than in the deep one, where undisturbed fresh sediment could cut off the old sediment from the lake interaction.

Applying the system of eqs. (6) and considering the present external P loading as constant (25t P yr^{-1}), the results are as follow:

$a = 0.2\ 10^{-3}\ m^3\ g^{-1}$	$a = 0.5\ 10^{-3}\ m^3\ g^{-1}$
$\alpha_1 = 1.51\ yr^{-1}$ $\alpha_2 = 0.11\ yr^{-1}$ ($\alpha_2{}^{-1} \sim 9$ yr) $P_{L\infty} = 130\ mg\ m^{-3}$ $P_{ads,\infty} = 0.83\ mg\ g^{-1}$ $P_{int,\infty} = 130\ mg\ m^{-3}$	$\alpha_1 = 1.34\ yr^{-1}$ $\alpha_2 = 0.09\ yr^{-1}$ ($\alpha_2{}^{-1} \sim 11$ yr) $P_{L\infty} = 126\ mg\ m^{-3}$ $P_{ads,\infty} = 0.84\ mg\ g^{-1}$ $P_{int,\infty} = 80\ mg\ m^{-3}$

Then, with the present P loading (~ 25 t P yr^{-1}) and the present state of the system (phosphorus saturated sediment and hypolimnetic oxygen depletion during a great part of the year), the future evolution of the Lake Varese system is foreseen as follows (fig.3):

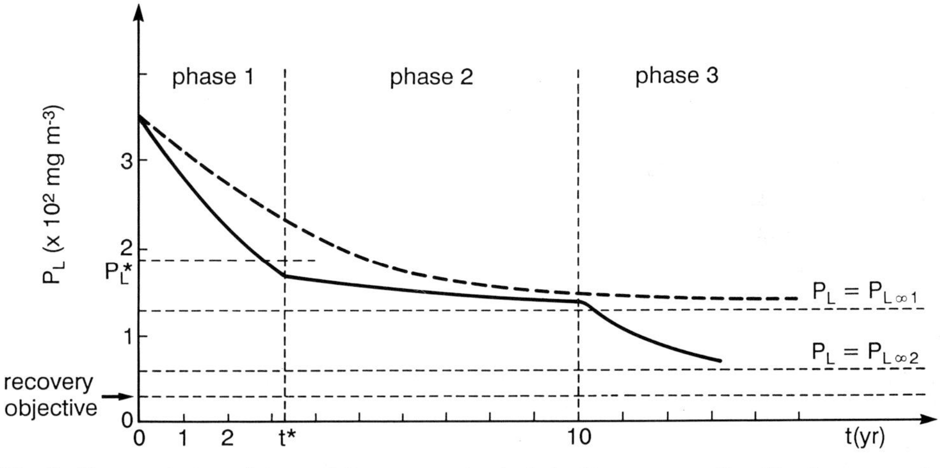

Fig. 3 Time evolution of the total P concentration in Lake Varese as predicted by the proposed model (t = 0 at 1987 winter circulation).

Phase 1: saturated P sediment (at t = t*: recovery of sediment structure starts).

Phase 2: progressive replacement of the old saturated layer with fresh unsaturated sediment (asymptotic P concentration in lake $P_{L\infty1}$ = 130 mg m^{-3}).

Phase 3: after reduction of external P loading to 12 t yr^{-1}, the new steady state concentration $P_{L\infty2}$ = 60 mg m^{-3} is "rapidly" attained (recovery objective = 34 mg Pm^{-3}).

The dashed curve gives the evolution with unsaturated sediment.

a) a relatively rapid (~3 yr) transition to partially unsaturated sediment, with a dynamics of dissolved P in pore water depending on the characteristics of the mineralization process, hence on the oxygen conditions at the interface water-sediment; and a sediment recovery time depending on the intensity of mixing processes in the active sediment layer;

b) a rather slow (~10 yr) approach to a steady state condition, with in-lake P concentration still to high to be acceptable (in reality phase a) and phase b) should be hardly distinguishable);

c) the simultaneous approach to uncoupling between lake and sediment (with increasing values of the distribution constant a) should lead to a lake dynamics progressively independent of the sediment evolution, hence rather rapid (characteristic time ~1-2 yr). Then the quality of lake water might rapidly respond to a further reduction of P load to attain an acceptable steady state P concentration. It has been calculated that a further 50% reduction of the present external P loads (to the goal of ~12.5 t P yr^{-1}, considered as a realistic one) might lower the total P concentration in lake water at a levels of 60 mg m^{-3}, approximately, after very few water overturns.

This is very important insight because the Water Management Plan of Lombardy Region has established, as realistic objective of (intermediate) lake recovery, a P concentration in water of 34 mg m^{-3}, but the model results suggest as ultimate (maximum attainable) goal of restoration a P level twice as high.

Sources and suggestions for further reading

1. E. De Fraja Frangipane, V. Mezzanotte and R. Vismara, The recovery of Lake Varese.
   Proceedings EWPCA Conference on Lake Pollution and Recovery, Rome, 1985, pp.391-394

2. G. Rossi and G. Premazzi, Delay in lake recovery caused by internal loading.
   Water Research, 25, 5, 567-575 (1991).
   (Drawings were reprinted from this paper with permission of Pergamon Press PLC.)

3. The Evolution of Copper Concentration in Lake Orta

3.1 Lake Orta and the history of its pollution

Lake Orta's pollution was originally caused by the effluents of a rayon factory, containing copper and ammonium sulphate, which were discharged into the lake without any previous treatment from 1926 to 1958 (maximum Cu input 80 t yr^{-1}).
Copper and ammonium have been treated in plants built inside the factory since 1958 and 1981 respectively, in such a way that the input of these pollutants diminished greatly after those dates.
An additional source of pollution has been created by several electroplating factories in Lake Orta's watershed, which have discharged into the lake considerable quantities of heavy metals (Cu, Cr, Zn, Ni) since the early 1970's, and this, together with the extreme acidity of lake water caused by in-lake ammonia oxidation (pH values as low as 4), has substantially slowed down lake recovery.
The morpho-hydrological characteristics of Lake Orta are summarized in tab. I and in the bathymetric map of fig. 1. The historical evolution of the average Cu and $N\text{-}NH_3+NH_4$ in-lake concentrations is presented in fig. 2.

Tab.I Morpho-hydrological data of Lake Orta

Lake level altitude 290 m a.s.l.
Surface 18.2 k2
Volume 1313 10^6 m^3
Watershed area 116 km^2
Maximum depth 144 m
Mean depth 71 m
Ouflow rate (average 1979-87) 4.57 m^3 s^{-1}
Theoretical water renewal time 8.9 yr

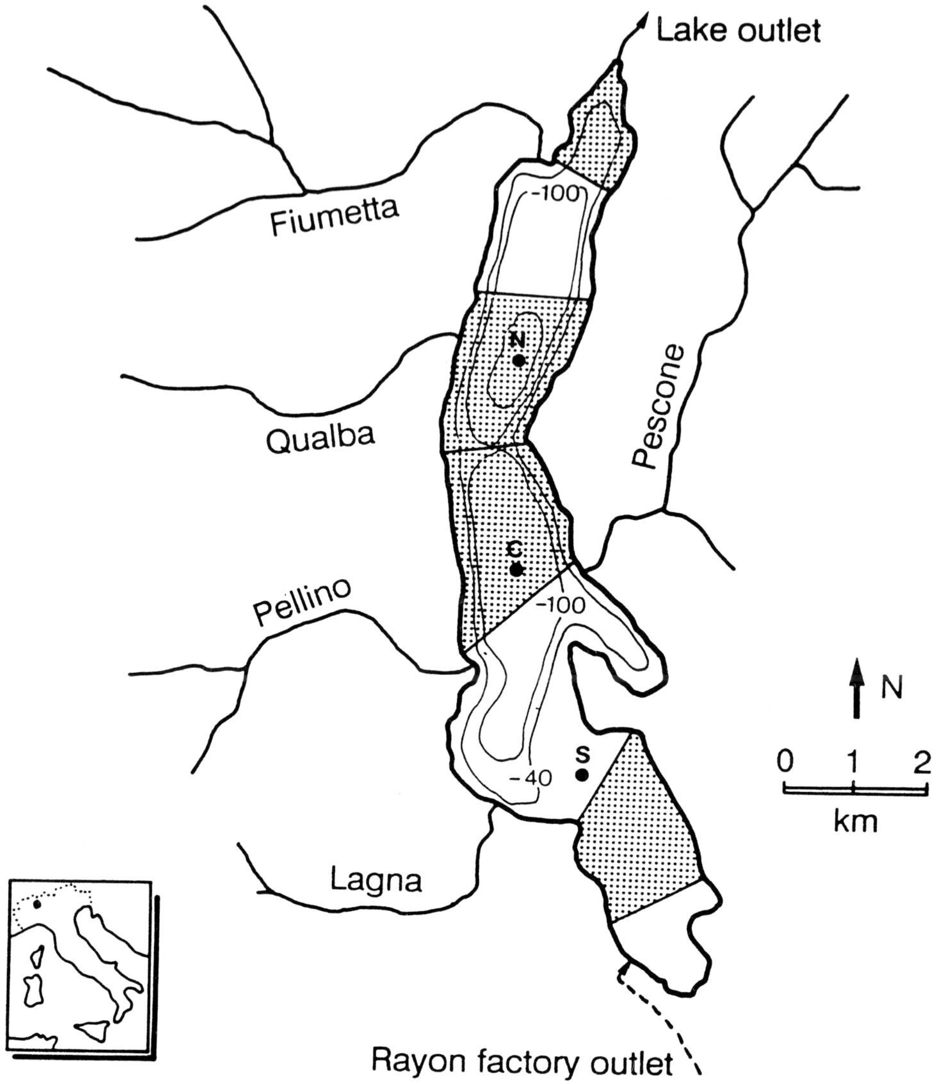

Fig. 1 Bathymetric map of Lake Orta (shaded areas: sampled lake sectors; S, C, N: sedimentation trap stations)

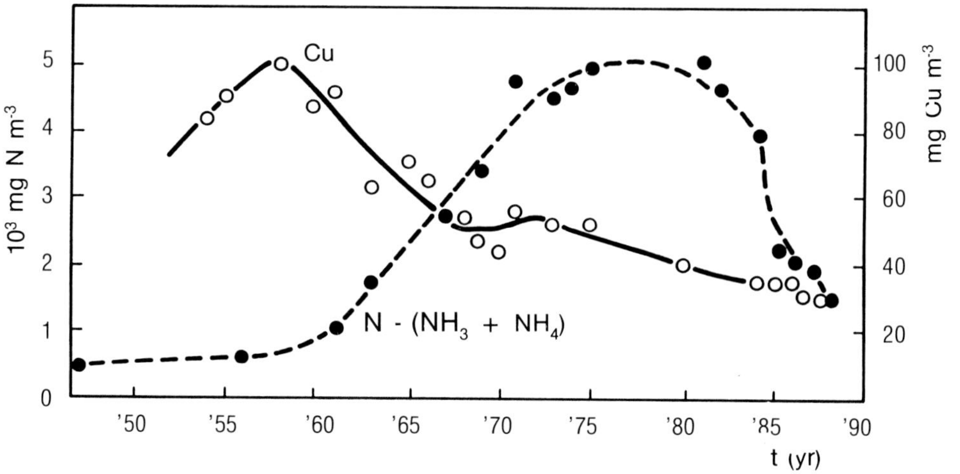

Fig. 2 Historical evolution of Cu and N-NH_3+NH_4 concentration in Lake Orta

3.2 The J.R.C. "Lake Orta Project"

In the "Water Quality" project of the programme "Protection of the Environment" at J.R.C., intensive experimental research was performed on Lake Orta from 1986 to 1989. Its aim was to ascertain the state of the lake, to assess the average external Cu load after diversion measures had been implemented and finally to predict the evolution of the Cu in-lake concentration towards an asymptotic level (corresponding to equilibrium with the assessed "residual" Cu load).
A "liming" experiment (water neutralization by addition of $CaCO_3$) was also begun under the responsibility of CNR/Istituto Italiano di Idrobiologia di Pallanza (IIIP) in the southern basin of the lake on May 1989, aiming at saturating the acidity caused by in-lake ammonia oxidation. Its effects have modified the properties of a considerable fraction of the water body under study to a greater or lesser extent. The sampling activity in the JRC project also covered the stratification period after the beginning of the experiment, but modelling considered essentially the pre-liming situation, leaving its long term effects to future investigation.

The long term predictive character of the study suggested the use of a metal distribution model with time scale in years, i.e. of total metal balance equations with time independent coefficients, which are accessible to analytical solution.
Such equations were written for the ecosystem considered both as a lake-sediment and a epilimnion-hypolimnion system, following the general concepts presented in Part II.

The analytical model was complemented by a computer code to solve the equations numerically, whenever the time dependence of the coefficients (e.g. the seasonal evolution of the parameters) had to be taken into account.
One of the basic conceptual assumptions of the modelling approach was that the interconnection between the two forms of Lake Orta pollution, copper and ammonium sulphate, could be summarized by the dependence of the copper distribution coefficient on pH.As a matter of fact the implementation of various pollutant diversion measures has not modified the extreme acidity of lake water significantly, and the effects of the recent "liming" operation are referred to a prescribed increase of the metal distribution coefficient in the epilimnion.

[The symbols used in the following sections are the same as defined in Part II]

3.3 Evaluation of the model parameters from experimental results

3.3.1 The mean residence time of water

The mean residence time of water in Lake Orta was evaluated by means of a computer code, on the basis of hydrological and thermometric data as specified below (see Part I, ch.3).

Water temperature profiles measured at the deepest point of the lake during 1984-1987 have suggested the following average annual evolution of the thermocline depth:

April, 1st-June, 15th	$z_{max}-z_{th}=16$ m	$V_{ep}=285.\ 10^6$ m^3
June, 16th-September, 15th	8.	147.
September, 16th-0ctober, 15th	12.	217.
October, 16th-November, 30th	16.	285.
December, 1st-January, 31st	20.	349.
Febr., 1st-March, 31st	143.	1313.

The "epilimnion" volumes V_{ep} were drawn from bathymetric data, via polynomial interpolation.

Hence Lake Orta was considered as holomictic and, on the basis of monthly average outflow rates at the lake outlet (Niguglia river) referred to the period 1979-1987, the results were as follows:

dilution of original water after 1 year: 91.1 %
water mean residence time: 10.7 years.

3.3.2 The vertical mixing coefficient

Only a rough description of the stratification process is required for the evaluation of the vertical mixing coefficient m_{th} to be used in long term predictions on the epilimnion-hypolimnion system.
On the other hand, calculations concerning the evolution of Cu concentration in the epilimnion during summer stratification should rely upon a careful evaluation of the mixing coefficient, such as it can be obtained by processing water temperature profiles, measured with sufficient space and time detail.
Unfortunately, such detailed information about water temperature was not included in the 1986-1989 JRC experimental campaign. Indications on the mixing coefficient at the depth of the thermocline, essentially in the form of average thermocline sinking velocity, have been drawn from sporadic (generally monthly) temperature measurements performed by IIIP and Istituto Ricerca sulle Acque (IRSA) between 1984 and 1989.
These velocities were found to be comprised between 0.04 and 0.09 m d^{-1}.
Then the following m_{th} values were adopted:

m_{th} (stratified lake) = 0.065 m d^{-1};
m_{th} (destratified lake) = 0.325 m d^{-1};
m_{th} (annual average) = 0.2 m d^{-1}.

3.3.3 The sedimentation flux of particulate matter

Sedimentation trap experiments were performed in Lake Orta under the responsibility of IRSA from 1987 to 1990. Settling material was collected at depths 30 m, 75 m, 120 m, 140 m in the northern basin; at depths 30 m, 75 m, 120 m in the central basin, at depth 30 m in the southern one (fig.l), on various dates to follow the seasonal evolution of particulate matter (tab. II).

In order to exclude the effects on particulate matter concentration of resuspension and $CaCO_3$ addition (liming experiment), the value of the sedimentation flux used in the model calculations (288 ± 29 g d.m. m^{-2} yr^{-1}) was obtained as an average of the fluxes which were measured at depth 120 m, 75 m, 30 m respectively in the three basins during 1987 and 1988. The mean deviation only takes into account the variability due to the different characteristics of the seasonal evolution in different years.

Tab. II Annual averages of sedimentation flux (g d.w. m^{-2} yr^{-1})

	1987	**1988**	**1988**	**1990**
Northern basin (–120 m)	228	190	207	318
Central basin (–75 m)	403	317	477	420
Southern basin (–30 m)	321	270	993	302

3.3.4 The Cu distribution coefficient

The results of the 1986-89 (pre-liming) sampling campaign are summarized in the mean values (and related mean deviations) presented in tab.III. The listed concentrations are "horizontal" averages concerning all sampling stations; "vertical" averages concern epilimnetic (–1 and –10 m) and hypolimnetic (–30, –110 and –130 m) depths separately; time averages concern 29 sampling dates between September 1986 and April 1989.

Tab. III Cu concentrations measured in Lake Orta 1986-89 (pre-liming)

Epilimnion **the whole year:**

C_p	=	1.232 ± 0.806 g m^{-3} (224 samples)
$(Cu)_p$	=	1.078 ± 1.010 mg g^{-1} corresponding to 1.328 mg m^{-3} (224 samples)
$(Cu)_{diss}$	=	35.4 ± 10.6 mg m^{-3} (224 samples)
$(Cu)_t$	=	36.75 mg m^{-3}
k_p	=	0.030 m^3 g^{-1} ($f_p = 0.972$)

Hypolimnion the whole year:

C_p	=	1.249 ± 0.878 g m^{-3} (108 samples)
$(Cu)_p$	=	0.953 ± 0.887 mg g^{-1} corresponding to 1.190 mg m^{-3} (108 samples)
$(Cu)_{diss}$	=	37.0 ± 9.7 mg m^{-3} (107 samples)
$(Cu)_t$	=	38.2 mg m^{-3}
k_p	=	0.026 m^3 g^{-1} ($f_p = 0.969$)

Epilimnion during stratification:

C_p	=	1.548 ± 0.876 g m^{-3} (80 samples)
$(Cu)_p$	=	1.389 ± 1.328 mg g^{-1} (2.150 mg m^{-3}) (79 samples)
$(Cu)_{diss}$	=	33.4 ± 11.5 mg m^{-3} (80 samples)
$(Cu)_t$	=	35.5 mg m^{-3}
k_p	=	0.042 m^3 g^{-1} ($f_p = 0.963$)

Hypolimnion during stratification:

C_p	=	1.507 ± 1.117 g m^{-3} (37 samples)
$(Cu)_p$	=	0.857 ± 0.414 mg g^{-1} (1.29 mg m^{-3}) (37 samples)
$(Cu)_{diss}$	=	36.0 ± 7.3 mg m^{-3} (37 samples)
$(Cu)_t$	=	37.3 mg m^{-3}
k_p	=	0.024 m^3 g^{-1}($f_p = 0.965$)

3.3.5 The mean residence time of copper in the lake

Whole lake averages are obtained from the epilimnetic and hypolimnetic concentrations of tab. III, being V_e = 147. 10^6 m^3; V_h = 1166. 10^6 m^{-3}: C_p = 1.247 g m^{-3}; $(Cu)_d$ = 36.8 mg m^{-3}; $(Cu)_p$ = 0.967 mg m^{-3}; $(Cu)_t$ = 37.8 mg m^{-3};
k_p = 0.0264 m^3 g^{-1} (f_p = 0.968)

With an effective mean residence time of water τ_w = 10.7 yr; a mean sedimentation flux φ_p = 288 g m^{-2} yr^{-1}; an average suspended matter concentration C_p = 1.247 g m^{-3}; a distribution coefficient k_p = 0.0264 m^3 g^{-1} and a mean depth h = 71 m, the mean residence time of Cu in Lake Orta was evaluated from the relationship:

$$1/\tau = 1/\tau_w + (1-f_p)(\varphi_p/C_p)/h = 1/\tau_w + [k_p/(1 + k_pC_p)](\varphi_p/h)$$
$$= 0.093 + 0.104 = 0.197\ yr^{-1}\ (\tau = 5.1\ yr).$$

3.4 Assessment of the residual Cu load

As for the external Cu load, the following (dissolved) Cu concentrations in inflowing waters were determined during the 1986-1989 sampling campaign:

Cu concentration in the precipitations: 11. ± 9.8 mg Cu m^{-3} (59 samples)
(precipitation rate: 1.9 m yr^{-1}x18.2 10^6 m^2 = 34.6 10^6 m^3 yr^{-1};
Cu input from precipitations: 0.38 t Cu yr^{-1};
Cu input from catchment area via infiltration: negligible);

Cu concentration in the coastline surface runoff: 103 mg Cu m^{-3} ("median" value from 27 samples)
(inflow rate from a suface 0.283 10^6 m^2 with infiltration coefficient 0.: 0.54.10^6 m^3 yr^{-1}
Cu input from surface runoff: 0.55 t Cu yr^{-1});

Cu concentration in the tributaries: 24.4 mg Cu m^{-3} (the mean value from 34 samples for each tributary was weighed on the inflow rate assessed via the area of the respective catchment surface with infiltration coefficient = 0.24; total inflow rate from the tributaries = 104 10^6 m^3 yr^{-1};
Cu input from the tributaries: 2.5 t Cu yr^{-1}).
It is worth mentioning that the concentrations measured in the ensemble of the tributaries, and even those measured in each river, present a very large dispersion because of the "non random" character of the waste.

As for the average contribution from the Bemberg factory after 1980, Bonacina and Bonomi (EAWPC Roma Conference 1985) account for a Cu input ~ 0.5 t Cu yr^{-1}.

Hence, directly assessed total external Cu load: 3.9 t Cu yr^{-1}.

3.4.1 Average Cu load compatible with the observed in-lake concentration decrease

In the framework of a simple mixed-reactor model, decrease ΔC of the average total Cu concentration during the time interval t_2–t_1, specific load J (t Cu m^{-3} yr^{-1}), and mean residence time of Cu in the lake are connected by the following relationship, which is obtained by integrating the differential balance equation:

$$\Delta C = [\tau J - C(t_1)]\,[1 - \exp(-(t_2 - t_1)/\tau)], \tag{1}$$

where:

$$\Delta C = C(t_2) - C(t_1)\,.$$

With reference to the total Cu concentration measured in the lake at the beginning of Seventies (a date after which the Cu concentration has decreased almost constantly), and to the results of the recent sampling campaign: $C(t_1) \sim 55$ mg m^{-3}; $C(t_2) \sim 38$ mg m^{-3}; $t_2 - t_1 \sim 15$ yr.

With the mean residence time assessed above ($1/\tau = 0.197$), the specific Cu load results $J = 7.27$ mg m^{-3} yr^{-1}, corresponding to an average annual load in the period 1973-88 $JV_L = 9.5$ t Cu yr^{-1}, and to an asymptotic total concentration $C_t = \tau J = 37.1$ mg Cu m^{-3}.

3.4.2 Evaluation of the average Cu load through Cu concentration in solid sediment

The relationship between total metal load I(t) (mg Cu m^{-3} yr^{-1}) and metal concentration C(z) (mg Cu g^{-1}) in the solid sediment as given in Part II, ch.4, in the assumption C = const(z) in the first few cm of the sediment core (eq. 7), applies to the case of Lake Orta during 1965-75 with C = 1.5 mg Cu g^{-1}.

Assuming for C_p, k_p the values recently assessed ($C_p = 1.247$ g m^{-3}; $k_p = 0.0264$ m^3 g^{-1}) in want of the historical ones, and $1/\tau = 0.197$ yr^{-1}, eq.(3) gives I = 11.6 mg Cu m^{-3} yr^{-1}, corresponding to a load of 15.2 t Cu yr^{-1} for that period.

The evolution post 1975 of lake and sediment has led to the Cu concentration in sediment C = 0.8 mg Cu g^{-1} as determined in the recent sampling campaign.

An attempt at evaluating the average Cu load from 1975 to 1990 through eq.(5) requires the use of the integrated equation (8) (with s = 0.001 m yr^{-1}, $z_o = 1.3$ cm). With the above quoted concentrations $C(z_o)$ and C(0), and assuming [C] = 1.15 mg Cu g^{-1}, eq.(8) gives [I] = 6.8 mg Cu m^{-3} yr^{-1}, corresponding to an average Cu load of 8.9 t Cu yr^{-1} for that period, in satisfactory agreement with the average Cu load evaluated on the basis of the observed Cu concentration decrease in the lake during the same period. Both results are about twice the external Cu load directly assessed during the 1986-1989 campaign. On the other hand, the results obtained with both methods are dependent on the sedimentation flux, the suspended matter concentration C_p and the distribution coefficient k_p, in such a way that their reliability must be discussed in terms of the reliability of the values adopted for these quantities.

3.4.3 The effect of the uncertainty in the measured parameters on Cu mean residence time in lake and assessed Cu load

It is assumed that the uncertainty in the mean residence time of Cu in the lake is essentially due to the uncertainty in the mean sedimentation time. In the assumption that $1/k_p >> C_p$, the latter is given by

$$1/\tau_{sed} = (\varphi_p/h)\, C_{ads}/C_d \qquad (1)$$

(all the symbols are defined in previous sections), in such a way that its mean deviation is expressed by means of the variances of the measured parameters φ_p, C_{ads}, C_d as follows:

$$D\,(1/\tau_{sed}) = \\ = (1/\overline{\tau}_{sed}) \left[D^2(\varphi_p)/\overline{\varphi}_p^{\,2}\right] + \left[D^2(C_{ads})/\overline{C}_{ads}^{\,2}\right] + \left[D^2(C_d)/\overline{C}_d^{\,2}\right], \qquad (2)$$

where D(x) and $D^2(x)$ are the mean deviation and the variance respectively and $\overline{x}$ is the mean value of x.
Mean values, mean deviations and variances of the measured parameters (C_d, C_{ads} referred to the hypolimnion) are found from the 1986-198 measurements (pre-liming) as given by:

$\overline{\varphi}_p = 288\ g\ m^{-2}\ yr^{-1}$; $D(\varphi_p) = 29$; $D^2(\varphi_p) = 841$

$\overline{C}_{ads} = 0.953\ mg\ g^{-1}$; $D(C_{ads}) = 0.887$; $D^2(C_{ads}) = 0.787$

$\overline{C}_d = 37\ mg\ m^{-3}$; $D(C_d) = 9.7$; $D^2(C_d) = 94.1$

The resulting mean deviation of $1/\tau_{sed}$ is ~ 0.101 (i.e. approximately equal to its mean value. The inverse of the mean residence time of water being $1/\tau_w = 0.093\ yr^{-1}$, the inverse of the total mean residence time of Cu in Lake Orta is given by $1/\tau = 0.197 \pm 0.101\ yr^{-1}$.
Note that the uncertainty in the characteristic time of the metal sedimentation process is due principally to the high variability of Cu concentration in particulate matter. This variability could be hardly reduced to purely stochastic variables. As an example of

the consequences of such high variability, the average Cu load compatible with the observed decrease of Cu concentration in the lake from 1975 to 1989, as estimated in section 3.4.1 (9.5 t yr^{-1}), is found to be affected by a relative error ~ 50%.

3.5 The evolution of Cu in-lake concentration towards equilibrium

If the values of the dissolved metal fraction f_p in epilimnion and hypolimnion on a annual basis are considerd (f_{pe} = 0.964; f_{ph} = 0.969, see tab. III), the only possible reason for a vertical heterogeneity in Cu concentration appears to be given by metal release from sediments.

As a matter of fact no real experimental evidence for such heterogeneity is supplied by the results (C_{et} = 36.7 ± 11 mg m^{-3}; C_{ht} = 38.2 ± 10 mg m^{-3}). Nevertheless, for the sake of completeness of the predictive information required, it seemed interesting to investigate the consequences on the long term evolution of the lake of the existence of a net metal release from sediments, covering a fraction of the assessed total Cu load.

In the presence of a net metal release flux from sediments φ_r (mg m^{-2} d^{-1}), the ratio $C_{et\infty}/C_{ht\infty}$ is given by eq.(5) of ch.3 in Part II.

The value 0.96 of this ratio (suggested by the experimental results) has been found to be consistent with

$$q_{in}C_{ext}/(\varphi_r S_{th}) \approx 10 .$$

Then a subdivision of the total Cu load (~ 9 t Cu yr^{-1}) into an external contribution (rivers, precipitation, runoff) of 8.2 t Cu yr^{-1} (0.017 mg Cu m^{-3} d^{-1}) and an internal one (net release from sediment) of 0.8 t Cu yr^{-1} (0.12 mg Cu m^{-2} d^{-1}) is proposed.

3.5.1 The evolution of the lake-sediment system

As for the lake-sediment interaction model, with the symbols defined above, the numerical values of the coefficients related to loading are as follows:

$k_{in}C_{ext}$ = 6.25 mg Cu m^{-3} yr^{-1}

pAv_rC_{Sd} = 800. 10^6 mg Cu yr^{-1}.

The values adopted for the lake parameters are discussed in section 3.3. In particular:

k_{out} = 0.093 yr^{-1} (inverse of water mean residence time);

k_{sed} = 3.253 yr^{-1}; k_p = 0.0264 m^3 g^{-1};

C_p = 1.247 g m^{-3} (f_p = 0.968).

The adopted values for the sedimentary quantities are as follows:

$p = 0.85$
$h_s = 0.03$ m
$\rho_s = 2.\ 10^6$ g d.m. m^{-3}.
$C_{Sd} = 40$ mg Cu m^{-3}; $C_{Sads} = 0.8$ mg Cu g^{-1}
$k_{ps} = 0.02\ m^3\ g^{-1}$
$r_s = \varphi_p = 288$ g d.m. $m^{-2}\ yr^{-1}$

With these values, release and settling velocities are:

$v_r = 0.8\ m\ yr^{-1}$; $v_s = 231\ m\ yr^{-1}$ (= $0.63\ m\ d^{-1}$).

The resulting asymptotic concentrations (eqs.4 and 5 of ch.2 in Part II) are as follows:

$C_{Lt\infty} = 33.6$ mg Cu m^{-3}; $C_{Sads\infty} = 0.77$ mg Cu g^{-1}.

The time constants of the system (eq.3 of ch.2 in Part II) are:

$\alpha_1 = 0.1994\ yr^{-1}$ ($\tau_1 = 5$ yr); $\alpha_2 = 0.0334\ yr^{-1}$ ($\tau_2 = 30$ yr)

The evolution of Cu in-lake concentration is almost exclusively determined by the residence time of Cu in the lake, the influence of the slowly decreasing release from sediments being negligible (under the square root of eq.3 in ch.2 in Part II, the "lake-sediment interaction" term is $1.57\ 10^{-3}$ against $26.\ 10^{-3}$).

3.5.2 The evolution of the epilimnion-hypolimnion system

A long term prediction of the evolution of the epilimnion-hypolimnion system can be obtained by means of the formalism presented in Part II, ch.3, the numerical values of the equation coefficients relying upon the lake parameter values as discussed in section 3.3.
In particular:

$k_{out,e} = q_{out}/V_e = 0.98\ yr^{-1}$
$k_{mix,e} = 8.9\ yr^{-1}$; $k_{sed,e} = 28.3\ yr^{-1}$; $f_{pe} = 0.964$;
$k_{mix,h} = 1.095\ yr^{-1}$; $k_{sed,h} = 3.57\ yr^{-1}$; $f_{ph} = 0.969$.

The source term in the epilimnion balance equation takes into account a constant external Cu load equal to 8.2 t Cu yr^{-1} ($k_{in,e}\ C_{ext} = 55.8\ mg\ m^{-3}\ yr^{-1}$) and that in the hypolimnion balance equation takes into account a constant Cu release from sediments equal to 0.8 t Cu yr^{-1} ($\varphi_r = 44.4\ mg\ m^{-2}\ yr^{-1}$).

The resulting time constants of the system (eq.2 of ch.3 in Part II) are:

$\alpha_1 = 11.6\ yr^{-1}$ ($\tau_1 = 0.086$ yr);
$\alpha_2 = 0.189\ yr^{-1}$ ($\tau_2 = 5.3$ yr).

As predicted in Part II, in the comments on the general formalism of ch.3, the long term evolution of the system is predominantly determined by the properties of the mixed lake.

The asymptotic Cu concentrations in epilimnion and hypolimnion (eqs.3 and 4 of ch.3 in Part II) are as follows:

$C_{et\infty}$ = 32.9 mg Cu m^{-3}
$C_{ht\infty}$ = 33.8 mg Cu m^{-3},

which give an average asymptotic in-lake concentration $C_{Lt\,\infty}$ = 33.7 mg m^{-3}.

The *numerical solution* of the model equations allows us to simulate the seasonal evolution of the mixing coefficient m_{th}, as well as that of epilimnetic parameters such as k_p, C_p.

As the source terms are the same as above (C_{ext} = 57 mg Cu m^{-3}; φ_r = 4 mg Cu m^{-2} $month^{-1}$), the calculations were performed with k_{pe} varying between 0.024 and 0.042 (m3 g^{-1}), C_p varying betwee 1.23 and 1.55 g m^{-3} in the epilimnion and beween 1.25 and 1.51 g m^{-3} in the hypolimnion and m_{th} varying between 2 and 10 (m $month^{-1}$). Because of the lack of historical information about the vertical heterogeneity of Cu concentration, initial (1975) concentrations in epilimnion and hypolimnion were both taken equal to 55 mg m^{-3}.

The resulting curves (see fig. 3) exhibit the approach to equilibrium concentrations, which are nearly uniform in the destratified lake at a level 32÷33 mg Cu m–3. The epilimnetic Cu concentration oscillates following the seasonal evolution of metal distribution between dissolved and adsorbed phase and thermocline stability.

3.6 The effect of the "liming" experiment on Cu distribution

The first monitoring campaigns after the liming experiments of 1989-1990 have shown a decrease of the epilimnetic Cu concentration from ~ 37 mg m^{-3} in April 1989 down to ~ 10 mg m^{-3} in August-September, an increase from 10 mg m^{-3} again up to 30 mg m^{-3} during winter 1989-90, and a new decrease down to 10 mg m^{-3} after the second part of the experiment (spring and summer 1990). The hypolimnetic Cu concentration has remained approximately constant at pre-liming levels, with tendency to a slight increase during summer 1990 (fig. 4).
As $CaCO_3$ was added at the end of the water total mixing period, its effect on the metal distribution concerned essentially a stratified lake. Because the evolution of water quality in a deep stratified lake is much slower in the hypolimnion than in the epilimnion, one is justified in assuming that, before a certain time, the variation of physico-chemical parameters caused by the experiment affects only the epilimnion, i.e. in assuming f_{ph} as unmodified during a certain time interval after modification of the epilimnetic distribution coefficient.

Then an interpretation of the experimental data as presented above was tentatively based on the system of balance equations (1) of ch.3 in Part II, with the values of the coefficients referred to the stratification period, i.e.:

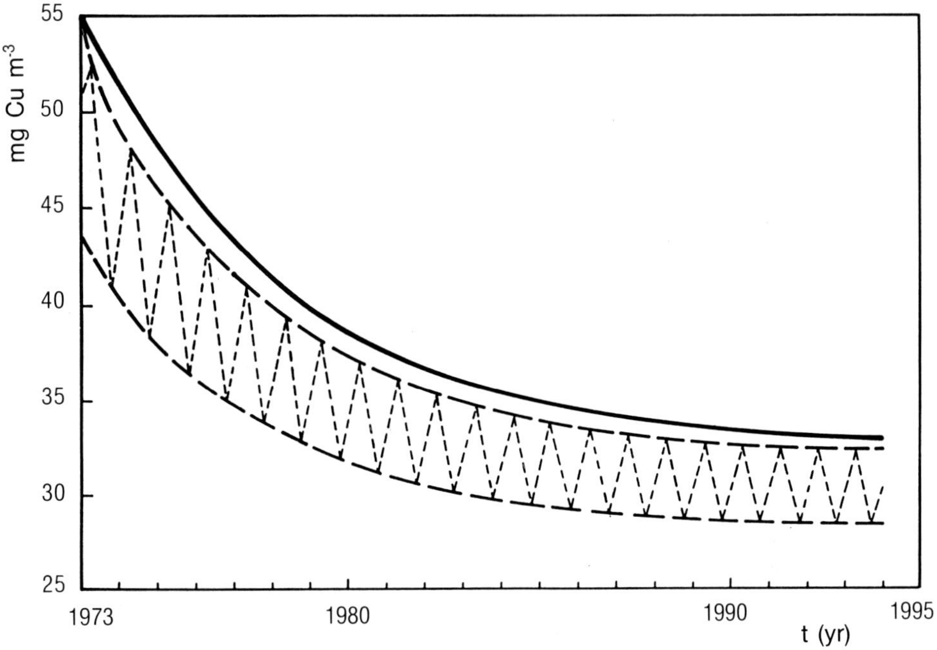

Fig. 3 Calculated evolution of Cu concentration in hypolimnion (——) and epilimnion (- - -) of Lake Orta

$k_{out,e} = 0.0027$; $k_{mix,e} = 0.008$; $k_{sed,e} = 0.08$;
$k_{mix,h} = 0.001$; $k_{sed,h} = 0.01$ (all in d^{-1})

With a total Cu load = 9 t Cu yr^{-1} (8.2 t external + 0.8 t internal):
$k_{in,e}C_{ext} = 0.153$ mg Cu m^{-3} d^{-1}; $\varphi_r/h_h = 0.0019$ mg m^{-3} d^{-1}.

As for the Cu distribution coefficient, the hypolimnetic value $k_{ph} = 0.024$ m^3 g^{-1} (f_{ph} = 0.965) was unmodified, while the epilimnetic value was tentatively chosen at the highest levels recorded in the surface layer of eutrophicated water bodies, i.e. k_{pe} = 0.150 m^3 g^{-1}. Furthermore, assuming an increase of the epilimnetic concentration of particulate matter to 2. g m^{-3}, f_{pe} is found to be decreased down to 0.77.

With these coefficient values and with initial values

$C_{et}(0) = 36$ mg m^{-3}; $C_{ht}(0) = 38$ mg m^{-3},

a short time approximation of the solution of the balance equations for total Cu in epilimnion and hypolimnion (i.e. a solution approximately valid during the first stratification period after the experiment), is obtained from the evolution curves by putting $\exp(-\alpha_2 t) \sim 1$:

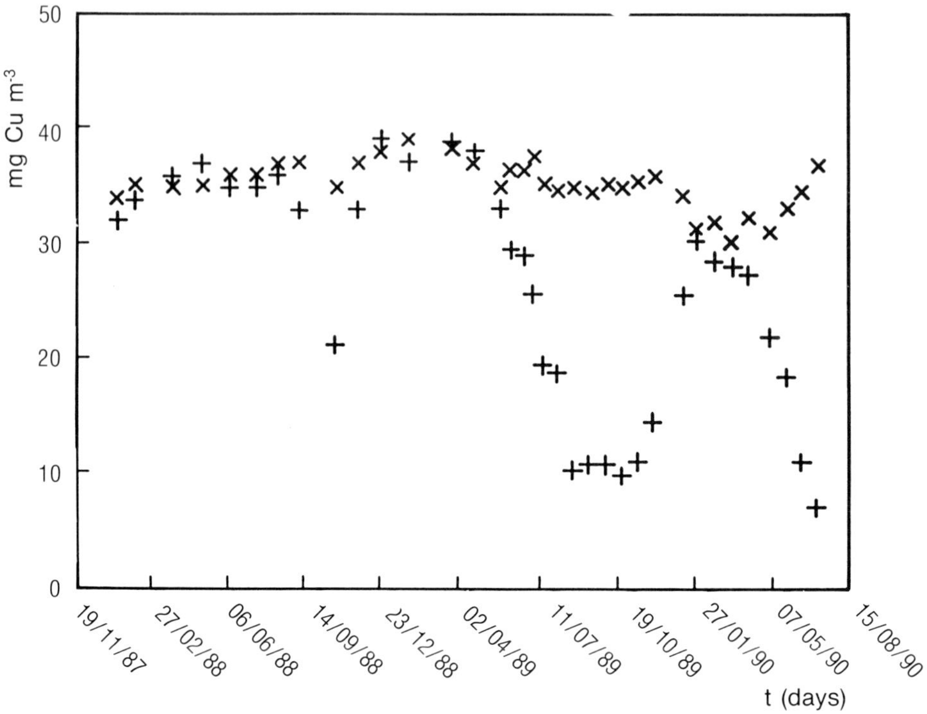

Fig. 4 Measured Cu concentration in epilimnion (+) and hypolimnion (x) of Lake Orta after liming experiments

$C_{et} \sim 19.5 \exp(-0.0282t) + 16.5$
$C_{ht} \sim -2.38 \exp(-0.0282t) + 40.5.$

These solutions, which obviously neglect the transition period with distribution coefficient varying from .026 to 0.150 m3 g–1 in the epilimnion, predict the following Cu concentrations after 90 d:

$C_{et} = 18$ mg m^{-3}; $C_{ht} = 40$ mg m^{-3},

a result which reproduces the experimental tendency in a satisfactory way.

3.7 Concluding remarks

The general behaviour of the calculated evolution curves (see in particular fig. 3) seems to indicate a simulated dynamics of the natural lake depuration process which is somewhat faster than the real one.

On the other hand, it was pointed out in section 3.4.3 that metal sedimentation and especially adsorption by particulate matter are taken into account on shaky experimental grounds (involving sampling and analytical procedures, which are difficult to improve).

The sensitivity of the evolution curves to the characteristic time of the sedimentation process is shown in fig. 5, where hypolimnetic Cu concentrations vs. time for Cu mean residence time τ = 5.1 yr (as assessed above) and τ = 5.8 yr (corresponding to 30% increase of the sedimentation time) are presented. Experimentally assessed average concentrations appear to be well located between the two curves.

As for the asymptotic Cu concentration, the practical interest of the model results, in the limits of the confidence which can be given to the adopted procedures to assess the residual external load, is to demonstrate that Cu concentration in Lake Orta would never reach acceptable levels, hence the great need for internal recovery measures, such as the neutralization of lake water acidity.

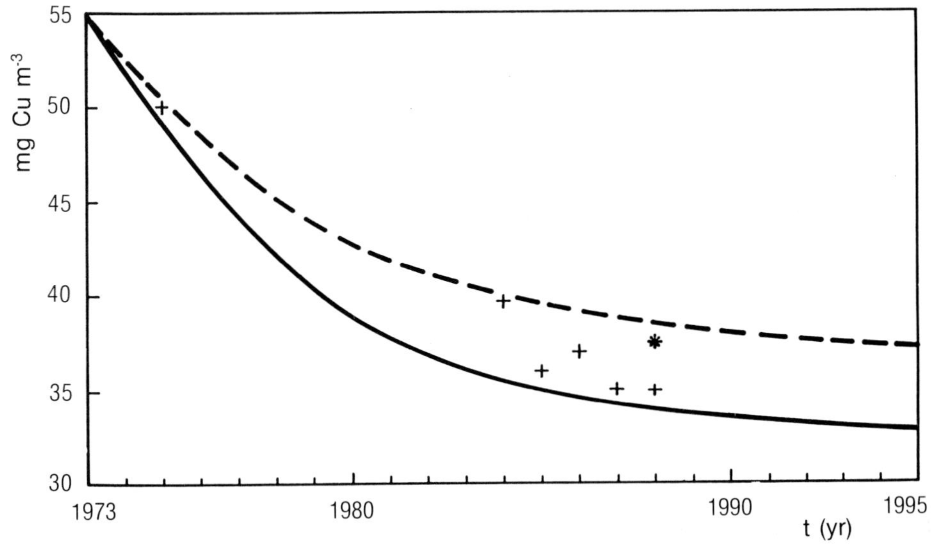

Fig. 5 Calculated evolution of average hypolimnetic Cu concentration in Lake Orta
—— Cu mean residence time 5.1 yr – – – Cu mean residence time 5.8 yr
+ exp. IIIP ✳ exp. JRC

Sources and suggestions for further reading

1. C. Bonacina, G. Bonomi and R. Mosello. Notes on the present recovery of Lake Orta: an acidic, industrially polluted, deep lake in North Italy.
   Mem. Ist. Ital. Idrobiol., 44 (1986), 97-115.

2. G. Rossi, H. Muntau and G. Tartari, The evolution of Cu concentration in Lake Orta.
   (to be published)

3. H. Muntau, G. Rossi, et al., Final Report "Lake Orta Metal Dispersion and Transport Study".
   EUR Report in press.
   (Figs. 2 and 4 are taken from: A. Calderoni, R. De Bernardi, R. Mosello, L'inquinamento del Lago d'Orta e suo risanamento mediante liming, Inquinamento, n. 11, nov. 1990)

European Communities - Commission

EUR 13998 — Modelling Lake Pollution

Giovanni Rossi

Luxembourg: Office for Official Publications of the European Communities

1991 - X, 144 pp. — 16.2 x 22.9 cm

Series: Environment and Quality of Life

EN

ISBN 92-826-3267-9

Catalogue number: CD-NA-13998-EN-C

Price (excluding VAT) in Luxembourg: ECU 12,50

Interdisciplinary studies on freshwater ecosystems, which were carried out at the Ispra Establishment of the Joint Research Centre of the Commission of the European Communities as activities of the project Water Quality (Programme Protection of the Environment), resulted in a series of scientific publications concerning various lakes of the subalpine region and various types of pollution.
Many of these studies followed a modelling approach, in the sense that field and laboratory experimental research was planned and results were interpreted with a mathematical model as a reference point.
This monograph is the result of a long interaction between modeller and experimentalist in tackling freshwater ecology problems. Its purpose is twofold: to present the concepts underlying lake modelling in a simple and concise way and to discuss a selection of lake pollution cases, as a synthesis of a pluriannual activity in which both interpretative and predictive modelling was developed.

After a short general introduction, Parts I and II are devoted to the modelling concepts currently used in the field of lake eutrophication and heavy metal pollution of lakes respectively.
In part III the application of such concepts to problems concerning three subalpine lakes is presented: the phosphorus cycle in Lake Lugano, the sediment-water interaction in Lake Varese and the evolution of the copper concentration in Lake Orta.

At the end of each chapter a list of titles is given as sources of additional information and suggestions to the reader for further development of the subjects concerned. An appendix to Part I contains a collection of parameter values adopted or suggested in recent publications on lake eutrophication modelling.

Venta y suscripciones • Salg og abonnement • Verkauf und Abonnement • Πωλήσεις και συνδρομές Sales and subscriptions • Vente et abonnements • Vendita e abbonamenti Verkoop en abonnementen • Venda e assinaturas

BELGIQUE / BELGIË

Moniteur belge / Belgisch Staatsblad

Rue de Louvain 42 / Leuvenseweg 42
1000 Bruxelles / 1000 Brussel
Tél. (02) 512 00 26
Fax 511 01 84
CCP / Postrekening 000-2005502-27

Autres distributeurs / Overige verkooppunten

Librairie européenne/ Europese Boekhandel

Avenue Albert Jonnart 50 /
Albert Jonnartlaan 50
1200 Bruxelles / 1200 Brussel
Tél. (02) 734 02 81
Fax 735 08 60

Jean De Lannoy

Avenue du Roi 202 /Koningslaan 202
1060 Bruxelles / 1060 Brussel
Tél. (02) 538 51 69
Télex 63220 UNBOOK B
Fax (02) 538 08 41

CREDOC

Rue de la Montagne 34 / Bergstraat 34
Bte 11 / Bus 11
1000 Bruxelles / 1000 Brussel

DANMARK

J. H. Schultz Information A/S

EF-Publikationer

Ottiliavej 18
2500 Valby
Tlf. 36 44 22 66
Fax 36 44 01 41
Girokonto 6 00 08 86

BR DEUTSCHLAND

Bundesanzeiger Verlag

Breite Straße
Postfach 10 80 06
5000 Köln 1
Tel. (02 21) 20 29-0
Fernschreiber:
ANZEIGER BONN 8 882 595
Fax 20 29 278

GREECE

G.C. Eleftheroudakis SA

nternational Bookstore
Nikis Street 4
10563 Athens
Tel. (01) 322 63 23
Telex 219410 ELEF
Fax 323 98 21

SPAÑA

oletín Oficial del Estado

rafalgar, 27
8010 Madrid
el. (91) 44 82 135

Mundi-Prensa Libros, S.A.

astelló, 37
8001 Madrid
el. (91) 431 33 99 (Libros)
431 32 22 (Suscripciones)
435 36 37 (Dirección)
élex 49370-MPLI-E
ax (91) 575 39 98

ucursal:

brería Internacional AEDOS

onsejo de Ciento, 391
8009 Barcelona
el. (93) 301 86 15
ax (93) 317 01 41

Llibreria de la Generalitat de Catalunya

Rambla dels Estudis , 118 (Palau Moja)
08002 Barcelona
Tel. (93) 302 68 35
302 64 62
Fax 302 12 99

FRANCE

Journal officiel
Service des publications des Communautés européennes

26, rue Desaix
75727 Paris Cedex 15
Tél. (1) 40 58 75 00
Fax (1) 40 58 75 74

IRELAND

Government Publications Sales Office

Sun Alliance House
Molesworth Street
Dublin 2
Tel. 71 03 09

or by post

Government Stationery Office

EEC Section

6th floor
Bishop Street
Dublin 8
Tel. 78 16 66
Fax 78 06 45

ITALIA

Licosa Spa

Via Benedetto Fortini, 120/10
Casella postale 552
50125 Firenze
Tel. (055) 64 54 15
Fax 64 12 57
Telex 570466 LICOSA I
CCP 343 509

Subagenti:

Libreria scientifica Lucio de Biasio - AEIOU

Via Meravigli, 16
20123 Milano
Tel. (02) 80 76 79

Herder Editrice e Libreria

Piazza Montecitorio, 117-120
00186 Roma
Tel. (06) 679 46 28/679 53 04

Libreria giuridica

Via XII Ottobre, 172/R
16121 Genova
Tel. (010) 59 56 93

GRAND-DUCHÉ DE LUXEMBOURG

Abonnements seulement
Subscriptions only
Nur für Abonnements

Messageries Paul Kraus

11, rue Christophe Plantin
2339 Luxembourg
Tél. 499 88 88
Télex 2515
Fax 499 88 84 44
CCP 49242-63

NEDERLAND

SDU Overheidsinformatie

Externe Fondsen
Postbus 20014
2500 EA 's-Gravenhage
Tel. (070) 37 89 911
Fax (070) 34 75 778

PORTUGAL

Imprensa Nacional

Casa da Moeda, EP
Rua D. Francisco Manuel de Melo, 5
P-1092 Lisboa Codex
Tel. (01) 69 34 14

Distribuidora de Livros Bertrand, Ld.ª

Grupo Bertrand, SA

Rua das Terras dos Vales, 4-A
Apartado 37
P-2700 Amadora Codex
Tel. (01) 49 59 050
Telex 15798 BERDIS
Fax 49 60 255

UNITED KINGDOM

HMSO Books (PC 16)

HMSO Publications Centre
51 Nine Elms Lane
London SW8 5DR
Tel. (071) 873 9090
Fax GP3 873 8463
Telex 29 71 138

Sub-agent:

Alan Armstrong Ltd

2 Arkwright Road
Reading, Berks RG2 0SQ
Tel. (0734) 75 18 55
Telex 849937 AAALTD G
Fax (0734) 75 51 64

ÖSTERREICH

Manz'sche Verlags- und Universitätsbuchhandlung

Kohlmarkt 16
1014 Wien
Tel. (0222) 531 61-0
Telex 11 25 00 BOX A
Fax (0222) 531 61-81

SVERIGE

BTJ

Box 200
22100 Lund
Tel. (046) 18 00 00
Fax (046) 18 01 25

SCHWEIZ / SUISSE / SVIZZERA

OSEC

Stampfenbachstraße 85
8035 Zürich
Tel. (01) 365 51 51
Fax (01) 365 54 11

MAGYARORSZÁG

Agroinform

Központ:

Budapest I., Attila út 93. H-1012

Levélcím:

Budapest, Pf.: 15 H-1253
Tel. 36 (1) 56 82 11
Telex (22) 4717 AGINF H-61

POLAND

Business Foundation

ul. Wspólna 1/3
PL-00-529 Warszawa
Tel. 48 (22) 21 99 93/21 84 20
Fax 48 (22) 28 05 49

YUGOSLAVIA

Privredni Vjesnik

Bulevar Lenjina 171/XIV
11070 - Beograd
Tel. 123 23 40

TÜRKIYE

Pres Dagitim Ticaret ve sanayi A.Ş.

Narlibahçe Sokak No. 15
Cağaloğlu
Istanbul
Tel. 512 01 90
Telex 23822 DSVO-TR

AUTRES PAYS
OTHER COUNTRIES
ANDERE LÄNDER

Office des publications officielles des Communautés européennes

2, rue Mercier
L-2985 Luxembourg
Tél. 49 92 81
Télex PUBOF LU 1324 b
Fax 48 85 73
CC bancaire BIL 8-109/6003/700

CANADA

Renouf Publishing Co. Ltd

Mail orders — Head Office:

1294 Algoma Road
Ottawa, Ontario K1B 3W8
Tel. (613) 741 43 33
Fax (613) 741 54 39
Telex 0534783

Ottawa Store:

61 Sparks Street
Tel. (613) 238 89 85

Toronto Store:

211 Yonge Street
Tel. (416) 363 31 71

UNITED STATES OF AMERICA

UNIPUB

4611-F Assembly Drive
Lanham, MD 20706-4391
Tel. Toll Free (800) 274 4888
Fax (301) 459 0056

AUSTRALIA

Hunter Publications

58A Gipps Street
Collingwood
Victoria 3066

JAPAN

Kinokuniya Company Ltd

17-7 Shinjuku 3-Chome
Shinjuku-ku
Tokyo 160-91
Tel. (03) 3439-0121

Journal Department

PO Box 55 Chitose
Tokyo 156
Tel. (03) 3439-0124